AF749272

Shelf 148·1

R

FORESTRY COMMISSION
WILDLIFE & CONSERVATION
RESEARCH BRANCH
LIBRARY

L/Wildlife
FORESTRY COMMISSION
ALICE HOLT LIBRARY.
No. 7185-78
Shelf. 148·1

BRITISH SNAKES

A. SMOOTH SNAKE (*Coronella austriaca*). The male and female are indistinguishable from each other in colour and markings.

B. Female GRASS SNAKE (*Natrix natrix helvetica*). The male is identical to the female in colour and markings, the only distinguishing feature being the length of the tail, which is longer in the male in proportion to the length of the body.

C. One male ADDER and two females basking together during the mating season. It is not unusual to find adders basking together in groups of three or more during the early spring. Later in the spring they separate into pairs. The smallest of the three is the male, which is black and greyish-white in colour. The female lying coiled is of a colour commonly seen, whereas the female lying extended is of a colour not often met with.

Leonard G. Appleby

BRITISH SNAKES

Illustrated by the author

JOHN BAKER

© 1971 Leonard G. Appleby
First published 1971
Reprinted 1975
JOHN BAKER (PUBLISHERS) LTD
4, 5 & 6 Soho Square
London W1V 6AD

ISBN 0 212 98393 8

All rights reserved. No part of this publication may be reproduced, stored in a retrieval system, or transmitted, in any form or by any means, electronic, mechanical, photocopying, recording or otherwise, without the prior permission of John Baker (Publishers) Ltd

Reproduced, printed and bound in Great Britain by MORRISON AND GIBB LTD, London and Edinburgh

Contents

List of Illustrations

TABLES (page 97)

Introduction

THERE have been very few books published which contain detailed information about the snakes indigenous to Britain. Bell's, *British Reptiles*, published in 1839, was for many years the only authoritative work on the reptiles and amphibians of this country. While it serves to identify the different species it tells us little about the habits and behaviour of British snakes.

In 1901 Dr Leighton produced the only book ever published that deals exclusively with the British species of snake. Although quite a thick volume a large part of it is taken up with letters from people living in different parts of the country describing the snakes that were to be found in their particular locality. As with Bell's book it contains little in the way of first-hand observations on snake behaviour. It was not until 1951 that a more informative work on the British reptiles and amphibians was published by Dr Malcolm Smith. As, however, the book covers both reptiles and amphibians it has not been possible to deal at any length with the snakes.

More than twenty-five years spent observing exclusively the three British species of snake has culminated in this book, and much of the information it contains has been obtained while I have waited many hours, and sometimes days, to photograph different aspects of snake behaviour. I have always felt inclined to agree with the adage that one picture is worth a thousand words, and for this reason I have endeavoured to obtain photographs of almost every form of behaviour known in snakes. I feel too that a photograph gives some proof that the naturalist who took it knows something of what he is writing about.

Unfortunately the snakes have never been such a popular subject for study as have most other members of the animal world. This may be due to the fact that many people abhor snakes, whilst some seem to have an almost inborn fear of them. Whatever the reason, the fact remains that there is much still to be learned about the habits and behaviour of these reptiles.

One of the most surprising discoveries in snake behaviour that has been made in fairly recent years is the spectacular displays of rivalry which occur between males of some species during the mating season. Fighting often takes place among the males of many species of snakes during the mating period and injuries are frequently inflicted. But in the displays of rivalry to which I refer no biting takes place; the two contestants simply fence or wrestle with each other to gain supremacy. This form of behaviour has been observed in only a few species, among them the rattlesnake of North America and some vipers, including the species indigenous to Britain. Some very good photographs of combat displays between male snakes are to be found in Clifford Pope's book *The Reptile World.* There are certain aspects of this type of behaviour which are not yet fully understood, and there is scope here for research which could result in even more surprising discoveries.

There are also other points of snake behaviour about which little is known, and one of these was, until recently, the type of food taken by the young of British snakes. It was always believed, and frequently published, that newly-born snakes fed on insects until they were able to tackle the type of prey consumed by the adult. The late Mr Jack Lester, who was for some years Curator of Reptiles at the London Zoo, shared my view that it was unlikely that baby snakes would feed upon insects, and he, like myself, was of the opinion that they fed upon the young of the animals they would devour when adult. By keeping infant snakes under close observation in captivity I discovered that, whereas insects and small invertebrates of all kinds were completely ignored, very small frogs, young newts, tadpoles, and baby lizards were devoured readily.

Snakes possess many attributes not shared by the majority of animals. These include the ability to identify scents and smells with the aid of the tongue, to swallow prey larger in girth than themselves, to live normally on one or two meals a week, and to survive several months if need be without any food at all. The snake can also climb without the aid of limbs, and swim without the assistance of either limbs or fins. In addition a snake can sleep with its eyes open, which it is forced to do since it has no eyelids, and, although it has no external ears and cannot hear in the way most animals do, it can detect by vibrations of the ground the approach of any large creature, such as man, many yards away.

It is because of the snake's extreme sensitivity to ground vibrations that many people have never seen a snake in its natural environment. To see snakes one should go alone, as two or more people searching together usually make far too much noise. Lightweight shoes and a cat-like tread are necessary if one wishes to observe snakes in the wild.

The snakes of this country are nowhere near as numerous as they were a hundred or so years ago. As long ago as 1945 Mr W. G. Wateridge, who was self-appointed snake-catcher for some sixty years or more in the New Forest, and who kept both the London and Edinburgh Zoos supplied with snakes, told me that snakes were less common in the New Forest then than they were in the late 1800s. The reader might feel inclined to retort that in view of Mr Wateridge's activities this was hardly surprising. But although Mr Wateridge may have slightly reduced the numbers of the New Forest snakes, the chief cause of their scarcity in many places where they were once abundant is the increase in the human population of Britain, and the consequent spread of residential areas. Large tracts of uncultivated countryside, which were once well removed from towns, now border large housing estates, and as a result many of the formerly unfrequented wild places beloved of snakes have become the recreation areas for people, causing the snakes to depart. For snakes do not like having their solitude disturbed, and they seem to abhor the close proximity of man almost as much as he abhors theirs. I refer of course to the general attitude towards snakes, although I am aware that there are some people who, like myself, actually like these reptiles.

Many heaths and commons which may appear ideally suited as habitats for snakes are often found to be entirely devoid of them, simply because they are too-frequented by people at weekends. With so many people now possessing their own transport there are few places which can now be described as being unfrequented. I know of several places in Surrey where snakes were once quite numerous, but the regular weekend invasion throughout the summer months of picnickers and ramblers has caused the snakes to disappear.

But is is not only the invasion by weekend explorers that has driven the snakes away from certain localities. The coming of World War II brought a more permanent invasion by the Army of large tracts of uncultivated land. In Surrey alone there are many once-beautiful heaths which became training grounds for soldiers during the last war, and which have never been relinquished. Some have become the sites of seemingly permanent army camps, while others have been retained as firing ranges.

But even if the army eventually vacates the heathlands which it at present occupies, it will probably be for the purpose of making room for more houses. It seems inevitable, as the human population of this country steadily increases, and more and still more new towns appear, that the countryside must slowly diminish, to the detriment of all wild life.

CHAPTER ONE

Identification

The three British snakes – length – growth – habitat – distribution – on the defence

WHILE there are approximately 2,700 different species of snakes to be found throughout the world, only three species are indigenous to Britain, these being, the Grass Snake (*Natrix natrix Helvetica*), the Smooth Snake (*Coronella Austriaca*), and the Adder (*Vipera berus*). The first two belong to the family, or group of snakes, named Colubridae, which is divided into different genera, the Grass Snake being a member of the genus *Natrix*, and the Smooth Snake of the genus *Coronella*. The Adder is included in the family *Viperidae*, and is of the genus *Vipera*.

Members of the genus *Natrix* are all non-venomous and semi-aquatic, feeding largely on cold-blooded prey, such as amphibians and fish, which is swallowed alive. The genus *Coronella* are also non-venomous, but have the ability to overpower and kill their prey by constriction before swallowing it, although they frequently swallow it alive without first constricting it. The *Vipera* are all venomous and use their venom to overpower prey, and also for defence. Both *Coronella* and *Vipera* feed on warm-blooded and cold-blooded prey.

A full description of each of the three British species is given in this chapter, together with details of their habitats, distribution, and defensive behaviour.

GRASS SNAKE (*Natrix natrix Helvetica*)

The grass snake does not exhibit much variation in its basic colour, the majority of specimens being some shade of green, usually darkish, while a few are olive-brown. When seen just prior to sloughing the grass snake appears a drab, dirty brown in colour, and its true colour cannot be correctly assessed until after the skin has been shed. Just behind the head there is a band of either yellow or orange which is usually, although not always, divided in the middle. Immediately behind this band there are two irregular shaped patches of black, which vary in size on different specimens. Grass snakes have been found whereon these black markings have extended for several inches along the sides of the body, but such cases are rare. The sides of the body are marked with short, vertical black stripes, which are narrow and straight on some snakes, and broad and irregular on others. Occasionally these stripes are edged with yellow or gold. Along the back there are two rows of black spots, sometimes replaced by short stripes. On some specimens the markings are absent from the back. The markings on aged grass snakes are often indistinct, and the colours are usually duller than on younger snakes.

The grass snake's underparts are usually an ivory white with transverse black bands, somewhat irregular in outline, which are sometimes fused together to form large patches of black. On some specimens the black markings form a chequered pattern, and occasionally they are completely absent. On one specimen I captured the black markings took the form of two black stripes running from head to tail, one on either side of the underparts, and separated by a band of white. The eye is golden or yellow in colour with a spherical black iris. The tongue is entirely black.

Albino grass snakes have been found, and J. Proctor (1926) records one kept at the London Zoo which was ivory-white, with orange eyes. The tongue was pink with a white tip, while the organs of the body were visible through the transparent underparts.

Completely black or melanic grass snakes have not been recorded in this country, although specimens have been captured in parts of Europe. Mr J. W. Lester, who was for some years Curator of Reptiles at the London Zoo, informed me that he had once received from Sweden a male and female grass snake both of which were completely black. This pair bred in captivity on two occasions, the first litter including some young that were jet black all over, and others that were not completely black. The second laying of eggs produced fewer truly melanic young ones than the first. On none of the young was there any sign of the yellow collar marking.

There is no colour variation between the sexes, and the markings are the same on both male and female. The only distinction between the sexes is in the length of the tail, which is proportionately longer in the male.

Length

The grass snake is the largest of the three British snakes, a fully grown female averaging 76 cm, the male some 15 cm or so shorter. It is not often that one finds a grass snake measuring more than 76 cm, and the largest I have captured measured 91·5 cm in length. I captured another specimen that was found to be 88 cm long, although it would have probably measured around 95 cm had not the last few centimetres of its tail been missing. Both snakes were females and were measured alive, being later released.

Much larger specimens have, however, been recorded in this country. Dr. Leighton mentions in his book, *British Serpents* (1901), a grass snake killed in South Wales in 1887 which measured 177·5 cm, and another, also taken in South South Wales, of 175 cm in length. This latter specimen is probably the same one as that referred to by Boulenger in his book, *Reptiles and Amphibians*, who also records another grass snake of 172·5 cm captured in the New Forest, and two of 160 cm taken in the Isle of Wight and Kent.

In 1943 Mr W. G. Wateridge wrote telling me about a grass snake measuring 189 cm long which he himself caught in the New Forest in about the year 1883.

Dr Leighton states in his book that he once captured a specimen 101·5 cm in length in North Monmouthshire, and there is an illustration in his book of a grass snake measuring

106·5 cm that was taken in Cornwall; whilst T. H. Gillespie refers in his book, *The Way of a Serpent*, to a grass snake of 114 cm in length as the largest he had ever seen from amongst hundreds of specimens sent to the Edinburgh Zoo during the years while he was director. He adds, however, that he handled a number of specimens measuring between 101·5 and 106·5 cm.

No really large grass snakes have been recorded since those giants of nearly two metres in length were found during the last century, and it seems improbable that a snake of such dimensions could exist undetected in this country today, in view of the enormous increase that has taken place in the human population.

While male grass snakes of just over 81 cm have been recorded the average length is around 66 cm. The female attains the greatest length, and any grass snake measuring more than 76 cm is almost certain to be a female.

Growth

Measuring approximately 18 cm at birth, the grass snake adds about 12 cm to its length in the first year of its life, and a further 10 cm by the end of the second year. The male attains sexual maturity when about three years old when it measures around 50 cm in length. The female becomes sexually mature in her fourth year and is then approximately 65 cm in length.

Habitat

Whilst showing a preference for clay soils the grass snake is frequently to be found on chalk downs, and also on peaty soils where the subsoil may be of either gravel or sand. Its favourite haunts are rough commons, where there is an abundance of long grass and bushes, and deciduous woods in which there are numerous large clearings where it can bask in the sun. Although often found in damp woods some distance from water, it is more numerous as a rule in the vicinity of a pond, stream, or river where the kind of animals upon which it preys are readily obtainable.

In localities where the subsoil is mainly of sand it is not unusual to find areas that are quite marshy, and in such places grass snakes are frequently to be found. Although it is commonly believed that the grass snake prefers a different environment to that chosen by the adder, the two species are often to be found in almost equal numbers in some places. I know of a Surrey common where one is as likely to meet with either an adder or a grass snake on a warm sunny day. The subsoil there is chiefly sand with a dark peaty topsoil, grass and heather grown intermingled, likewise both deciduous trees and conifers. Parts of the common are damp and almost marshy even during the hottest months of the year, while in other parts the ground is quite dry and dusty during the summer months, and one is equally likely to find a grass snake in one of the dryer areas of the common and the adder on the marshy ground.

That the grass snake is not altogether averse to dry situations is borne out by the fact that I once found a mating pair on top of Coneyhurst Hill in Surrey, where the subsoil is of sand, and far removed from water in any form.

Distribution

Within the British Isles the grass snake's range extends from Cornwall to the east coast, and as far north as the southern borders of Cumberland and Durham. It is much less common north of the Midlands than in the southern half of England, where it is numerous. It is also to be found in Wales, and the Isle of Wight, but is completely absent from Ireland. Outside the British Isles its distribution covers France, Western Germany, Switzerland and Italy.

Defensive Behaviour

Like the animals it preys upon the grass snake has very little defence against its enemies. It usually seeks to escape when molested, and if caught and held it invariably releases from the vent a foul-smelling white fluid, which is a mixture of excreta and anal gland secretion. Whether or not this would in any way deter a predatory animal intent on making a meal of the snake is doubtful. It is possible that a predator, having killed the snake, might be disinclined to devour it because of the unpleasant smell, but this would be of little advantage to the snake once it was dead. However, if the snake could pretend to be dead after emitting its strong odour a predator might take no further interest in it. This is, perhaps, the reason why grass snakes sometimes feign death when attacked or captured.

When first caught the grass snake usually struggles violently to escape, at the same time hissing loudly, and making feinting strikes with its head. If unable to get away, the secretion just referred to is voided, and if this fails to secure the snake's release it may feign death. If on the ground it does this by rolling over on to its back, opening its jaws wide, and protruding the tongue with the bifid tip closed. The snake lies quite still just as if dead, but if turned on to its belly it will often roll over on to its back again. Not all grass snakes insist on lying upside down when feigning death, and some will lie in whatever position they are placed, but invariably with jaws agape, and with the tongue hanging loosely from the mouth.

If lifted from the ground when captured the grass snake, after struggling violently, feigns death by suddenly becoming limp, and hangs loosely from its captor's hand. I have carried a death-feigning grass snake for over half a mile without its showing the slightest sign of life. If replaced on the ground, or if left lying on the ground in the first instance, the death-feigning grass snake will come to life again within a minute or two and glide quietly away, providing the observer keeps quite still. But if picked up just as it is making off the snake may feign death a second time, but not always. Many grass snakes will sham death only once, and thereafter only struggle to escape. Once removed from its natural environment and taken into captivity the grass snake can seldom be induced to feign death, although it may struggle to escape, and release its anal fluid as before.

When cornered by an aggressor and unable to escape the grass snake will try to bluff its way out of trouble by adopting a threatening posture. Drawing up the forepart of its body into an S-shaped loop, it inflates itself with air which it expels with an explosive hiss, at the same time lunging forward with its head in a manner reminiscent of the adder when striking. The grass snake's jaws remain closed when it strikes and its head seldom makes contact with the object at which it is striking. In between these feinting strikes the tongue is flickered nervously in and out between closed jaws. If the aggressor keeps his distance and remains still the snake ceases its angry hissing and oscillates its tongue slowly as though trying hard to identify its enemy by scent. Sometimes only the bifid tip of the tongue is protruded and held stiffly in this position for a while. Whether or not this is intended as a threatening gesture I am not certain. The impression gained is that the grass snake is trying to fool its enemy into believing that the stiffly held tongue is some kind of sting, but this may not be the intention.

In my experience the grass snake rarely bites, or makes any attempt to bite. Its threatening behaviour when cornered is chiefly bluff, although this can be very convincing. I have more than once seen a dog kept at bay by a grass snake indulging in this kind of behaviour, and no doubt it would be equally effective against some wild animals.

Only once have I known a grass snake make an attempt to bite in self-defence, and that happened while I was trying to photograph the snake. I had cupped my hand over the coiled snake to restrain its efforts to escape, and as I lifted my hand the snake struck at it in the manner of an adder, its jaws opening wide as it lunged forward, and jabbed its tiny recurved teeth into the palm of my hand. I felt the teeth catch the skin on my palm, but on examination immediately afterwards I could find no marks on the skin to show where the teeth had made contact.

Such behaviour is unusual for the grass snake, and in all

the many years I have spent studying this snake it is the only instance I have recorded of a grass snake biting. The fact that the grass snake in question was at the time carrying eggs, which it laid shortly afterwards, may have accounted for its abnormal behaviour on that occasion.

Smooth Snake (*Coronella Austriaca*)

One has only to hold this snake in the hand to realise why it has been given the name smooth snake, for like the slow-worm it is smooth and polished to the touch. The reason for this is the absence of the central ridge or keel which is present on the dorsal scales of both adder and grass snake.

There is no colour variation between male and female smooth snake, and the only means of identifying the sexes is in the length of the tail, which is slightly longer in the male, being approximately one-fifth of the total length, and one-sixth of the total length in the female.

Although sexual variation in colour is lacking there is some variety in the colours and markings of different specimens. I have seen brown, reddish-brown, straw-coloured, grey, and olive-grey smooth snakes. The ventral surface colours also vary from one individual to another. In grey or olive-grey specimens it can be either grey or almost black, whilst reddish-brown smooth snakes may have the underparts a uniform pinkish or red colour, or a creamy yellow with either pink or red spots and markings. Brown and straw-coloured specimens are usually a lighter shade of the dorsal colour on the under-surface. In all smooth snakes the underside of the jaws and tail are invariably a lighter shade of the belly colour.

On the back of the neck there is a small transverse dark band or irregular shaped patch. On some specimens the markings on the body are confined to a double row of spots along the back, sometimes joined together to form short bars. On others there is also present another row of spots, not always clearly defined, along the sides of the body. A distinctive marking which is always present is the dark line on either side of the head, which extends from the nostrils, through the eye, to the side of the neck. The markings on the smooth snake are usually a darker shade of the basic colour, and on reddish-brown specimens the spots are often a dark red in colour.

The eyes of the smooth snake can be either amber or yellow, the pupil being black and spherical like that of the grass snake.

The tongue of this snake differs from that of both grass snake and adder in that it is often a chocolate brown in colour, fading to a pinkish hue at the base. It is normally the darker specimens, such as grey and olive-grey, that possess a black tongue, although this is sometimes brown at the base.

Length

There is no difference in length between male and female smooth snakes, and the average length for a fully grown specimen of either sex is about 50 cm, although larger specimens of up to 63 cm have been found. Specimens in excess of this length have not been recorded in Britain.

Growth

At birth both male and female average 17 cm in length, and by the end of the second year of life this will have been doubled. Sexual maturity is reached at the age of four years when both sexes will measure approximately 45 cm in length.

Habitat

Sandy heathlands where there is an abundance of heather are the favourite haunt of this snake, and it is more likely to be found in the vicinity of a pond, stream, or marshy ground than in areas far removed from water. It also inhabits open wooded slopes with a southern aspect. In common with the adder, the smooth snake feeds chiefly on lizards and consequently its choice of habitat is very similar, although its range is much more restricted.

Distribution

The distribution of the smooth snake within the British Isles is limited, as far as is known, to the three southern counties of Surrey, Hampshire, and Dorset, and although not particularly abundant in any of them it is, perhaps, more common in Hampshire than in either of the other two counties. In Hampshire it is found mainly within the area covered by the New Forest, whilst in Surrey the heathlands in the neighbourhood of Hindhead and Churt are the places where it has most frequently been found.

Outside the British Isles the smooth snake's range covers France, Germany, Austria, Holland, Norway, Sweden, Switzerland, North Spain, North Portugal, Albania, Macedonia, and the northern part of Greece.

Defensive Behaviour

Although non-venomous the smooth snake can, and often does, defend itself by biting. Unlike the adder and the grass snake it never attempts to bluff its way out of danger by hissing or inflating its body with air. It is altogether a much calmer snake by nature than either of the other two British species, seldom becoming excited or angry, and it rarely hisses at all. When it bites it does so slowly and deliberately, and completely without warning. It bites hard, moving its jaws forward on to the flesh of its aggressor in the same manner adopted when swallowing prey. Although its teeth are small they are needle sharp, and long enough to pierce soft skin and draw blood. Once having taken hold it is reluctant to let go.

Most smooth snakes are inclined to bite when first captured, but they soon become accustomed to being handled, and after a short while in captivity they can usually be handled without their attempting to bite, unless roughly treated.

During the mating period male smooth snakes frequently fight amongst themselves for possession of a female, and during these combats they will bite one another quite viciously. But once mating is finished there is no sign of animosity between the males, and in captivity they will bask amicably enough together. However, this snake is easily provoked, and I have seen one smooth snake bite another because it was disturbed while basking by a companion that was feeding, and which hitched its tail on to the basking snake. The feeding snake continued with its meal as though nothing was amiss, and made no attempt to retaliate.

On another occasion I was removing a smooth snake from a bag containing several when it hitched its tail around another smooth snake, which resulted in the two being lifted out together. The second snake obviously objected to being hoisted aloft in this manner and promptly bit the other, chewing its jaws hard into its body. The bitten snake did not retaliate, and on examining it afterwards I could find no trace of teeth marks, although a few days later I noticed a slight blackening of the scales where the bite had been received.

Occasionally the smooth snake will discharge a strong smelling fluid from the vent when first captured, and I have known it to do this in captivity when handled. This form of behaviour is, however, unusual in the smooth snake, and I have only known a few to react in that way when handled.

Adder (*Vipera berus*)

Of the three British snakes the adder exhibits by far the greatest variation in colour. It is also the only British species in which the sexes can be identified by their colour.

When first born the female adder is usually a dull brown, but during the first year of life many become brick red in colour and remain so until they are about 25 cm in length. As they increase in size beyond this length their colour gradually changes again, so that by the time they reach maturity they will have become golden brown or brown. It was at one time believed that these immature red female adders were a species distinct from Vipera berus. Bell, in his book *British Reptiles* (1839) gives it the title, Small Red Viper, and Dr Leighton in his *British Serpents* (1901) devotes a whole chapter to it, and he was equally convinced that it was a species different from that of the larger specimens he had examined.

The male adder also begins life a dullish brown in colour which slowly changes to a straw yellow as the snake grows, until finally it becomes grey or dull white on reaching maturity.

The ventral surface of mature specimens of both sexes may be brown, grey, light blue, gun-metal blue, or black, the underside of the tail being a slightly lighter shade, whilst the throat is usually white, yellowish, or orange, sometimes marked with either black or brown. The colour of the eye is orange-brown or copper in females, whilst in the male it is usually a deep garnet red. The iris or pupil of the eye is black and takes the shape of a vertical elliptic, which narrows to a fine slit in bright sunlight, and widens progressively in a diminishing light until it becomes completely round in shape. This type of eye usually denotes nocturnal habits, but in the case of the adder this is not true, since the reptile's chief functions, such as mating and feeding, are carried out in daylight, most usually bright sunlight, although the adder does sometimes feed at dusk during very warm weather. The adder's tongue, like that of the grass snake, is entirely black.

Bell mentions in his book *British Reptiles* (1839) that the skin of the adder is iridescent when viewed in certain lights. I too have observed this, and most frequently in specimens that have recently sloughed. I first observed this iridescence one summer some years ago whilst watching a basking female, the scales of which appeared to possess a greenish sheen when looked at from a particular angle.

On the back of the adder's head there is a distinctive marking that resembles an inverted letter V. This marking varies slightly in different adders, and sometimes it more closely resembles the letter H with the upright strokes splayed outwards at the base. A broad zig-zag band extends along the full length of the back, while along the sides of the body there is a single row of spots which diminish in size towards the head and tail. The zig-zag band is broader and the spots larger in the male than in the female. These markings are a darker shade of the basic colour in the female, while in the male they are usually jet black, sometimes edged with brown. Occasionally the zig-zag marking along the back is replaced by a straight band. Bell illustrates in his book, *British Reptiles*, an adder with this type of marking, and a photograph of a similarly marked specimen appears in Dr Malcolm Smith's book, *British Amphibians and Reptiles* (1951).

While an albino adder has never been recorded in this country, black or melanic specimens have frequently been found. Truly melanistic adders, meaning those which are completely black with no trace of markings, often breed true, producing entirely black young. As a result of this black adders are frequently found in certain localities, and are completely absent from others. An instance where a black adder did not breed true was brought to my attention by the well-known herpetologist Alfred Leutscher, who wrote to me in 1958 about a black adder he once found in the New Forest, and which was 'a shiny jet black with no trace of markings, just a slight reddish tinge around the lower jaw. She gave birth to five normally coloured and marked young when I got her home'.

Adders of unusual colouring are sometimes to be found. I obtained from South Devon a large female 61 cm in length that was a greenish-grey in colour. It was quite the most unusually coloured specimen I have ever seen.

Length

The female adder grows to a larger size than the male and averages about 55 cm in length, the average male measuring some 5 cm less, although slightly larger specimens of both sexes are not uncommon. The largest adder I have ever captured was a female measuring 61 cm in length and was taken on Arbrook Common in Surrey, whilst the largest I have handled was a female 66 cm in length that was sent to me from South Devon. It looked a veritable giant in comparison with more normal sized specimens. I also obtained from Devon a male measuring 61 cm.

Adders greater in length have been found, and it is recorded in the Annual Report of Christ's Hospital (Horsham) Natural History Society than an adder measuring 85 cm was caught by a Mr R. Robertson in St Leonard's Forest in Sussex, on 16 July 1926. The second largest adder recorded was that taken in Herefordshire by Dr Leighton and which measured 72 cm. A female measuring 82·5 cm and a male of 72·5 cm in length are claimed to have been caught by a Mr W. G. Wateridge in the New Forest some time around 1880, but there is no authentic record of these two specimens.

Growth

At birth both male and female average 17 cm. In the male this length is increased by 10 cm in the first year of life, a further 8 cm being added by the end of the second year, and by the time it is four years old and sexually mature it measures about 40 cm in length. The female grows at a slightly faster rate, and although sexually mature at the same age she measures approximately 47 cm in length at the end of her fourth year.

Habitat

Heathlands and commons, where the subsoil is of sand, are the favourite haunts of the adder. It is seldom found on clay soils, but it is frequently met with on chalk downs, and in some parts of the South Downs it is numerous. It is also fairly abundant on the southern side of the North Downs, although not often found on the colder, northern side. Open woodland, uncultivated fields, disused railway cuttings, unworked quarries, sandpits, chalk pits, and even deserted airfields, are also places favoured by the adder. It is less often found near human habitations than the grass snake, preferring places unfrequented by people.

Feeding mainly as it does on lizards and mice, both of

which prefer dry situations, the adder does not require to be in the neighbourhood of water. However, it has been my experience that adders are to be found more abundantly in the vicinity of a pond or stream than in the very driest of localities. The fact that it likes to drink as frequently and as copiously as the grass snake may account for this. The adder certainly seems in no way averse to dampness, as I have found it inhabiting low-lying areas that were often waterlogged during the winter and spring, and only really dry during the hottest part of the year. One of these was Arbrook Common in Surrey, which was at one time a favourite haunt of adders. The southern end of the common was always wet and spongy underfoot during the spring months, and yet adders could often be found there at that time.

Adders are also numerous on the boggy ground in the vicinity of Pudmore Pond on Elstead Common in Surrey. Although referred to on the Ordnance Survey map as a pond there is in fact little open water, except in the very centre. Pudmore Pond consists mainly of marshland covering an area of roughly 600 square yards. Heather grows abundantly on the fringes of the pond and also covers most of Elstead Common, much of which is waterlogged even during the summer months. The water there is not the direct result of rainfall but rises from a number of small springs which are dotted about the low-lying areas of the common.

My first acquaintance with Elstead Common was in the spring of 1947, when I found an adder basking on a large tuft of grass on the edge of Pudmore Pond. There were similar grass tussocks all around, and all of them were surrounded by water some 10 to 15 cm in depth. When I found the adder I was making my way towards the centre of the pond, stepping from one grass tussock to the next in order to avoid getting my feet wet. The nearest dry ground was a sandy bank about ten yards away which marked the edge of the pond.

Subsequent visits to Elstead Common revealed that adders were abundant there, and I rarely visited the place without seeing at least one adder, usually more. They were to be found chiefly on the dry banks of small, water-filled canals, and on small islands of dry ground raised above the surrounding waterlogged, heather covered common. To get from one dry patch of ground to another the adders would have no choice but to traverse the boggy ground, and they would need to travel some distance at times, for the dry islands are few and far between. That the adders should choose to inhabit this decidedly damp place when there are larger areas of higher and much drier ground to be found on other parts of the common is contradictory to the accepted behaviour of this reptile. Possibly they have discovered that they are less likely to be disturbed by intruders there, for few people venture on to the marshy ground. This is not because it presents any danger, but it is deceptive, and areas which appear dry from a little distance away are often found to be quite waterlogged, and one can suddenly and unexpectedly find oneself floundering about almost knee-deep in boggy ground.

W. H. Hudson, who was particularly fond of adders, and who obviously knew their habits, refers in his *Book of a Naturalist* to adders inhabiting 'an extensive boggy heath'.

That the adder is not averse to entering water I have discovered with captive specimens. I have known adders to lie in water when the temperature of it has been above that of the surrounding air. I have also watched a captive specimen take a drink from the water receptacle or pool in its enclosure and then enter the water and swim about beneath the surface.

Distribution

The adder is found throughout the British Isles, with the exception of Ireland, and is equally as numerous in the north of the country as it is in the south. Outside the British Isles its range extends over most of Europe, from northern Spain and northern Italy to Scandinavia and Finland. It also inhabits parts of northern Asia and eastern China.

Defensive Behaviour

A detailed account of the adder's defensive behaviour is given in chapter 9.

CHAPTER TWO

Anatomy, locomotion, growth, intelligence and enemies

Snakes and legless lizards – anatomy – means of locomotion – uses for tongue – yawning – intelligence – dangers

ONE obvious and distinguishing characteristic of the snake is its lack of limbs. But it does not follow that all reptiles without limbs are snakes, as there are many species of lizard which are also legless, and some are very snake-like in appearance. Lizards, however, possess certain characteristics which no snake owns, the most obvious of these being movable eyelids. Examination of a legless lizard will also reveal that the scales covering the underside of the body are similar to those on the back. On the underside of a snake there is only a single row of broad scales, known as the ventrals, extending from the neck to the vent. The scale covering the vent is called the anal shield and is divided in the middle, while those covering the undersurface of the tail are known as the subcaudals, and are also divided. The back is covered with several rows of small leaf-shaped scales termed the dorsals.

In most snakes the head is clearly defined from the body by a tapered neck, whereas on a legless lizard the head joins the body with no noticeable division between the two. Furthermore, the snake has loosely hinged jaw bones which allow of considerable expansion of the jaws to permit large meals to be swallowed whole. The lizard's jaws are rigidly hinged and only small creatures, no larger in girth than the lizard's head, can be swallowed.

Both lizards and snakes possess a forked or bifid tongue, but there is a difference in the form of this organ between the two reptiles, that of the lizard being blunter and notched at the end, whilst the snake's tongue separates into two very finely tapered threads. To protrude the tongue the lizard has to open its mouth, whereas the snake is provided with a special notch or indentation at the front of the upper jaw through which the tongue can be extended with the jaws closed.

Anatomy

When, during the slow process of evolution, the snake discarded its limbs and became elongated and slimmer in form, its internal organs also underwent a change. With the narrowing of the body there was less room for two lungs side by side to function efficiently. Consequently, the left lung shrivelled and became a vestigial organ in most snakes, although a few primitive species still possess two lungs. The single lung possessed by most snakes is only a simple affair, and is little more than a reservoir for air.

The snake's heart is equally unsophisticated, having only three chambers, which include a left and right auricle, but only a single ventricle. As the ventricle is incompletely partitioned, the oxygenated blood from the lungs tends to get mixed with de-oxygenated blood returning from the body tissues. This imperfect arrangement, plus the inability of the lung to oxygenate the blood efficiently, makes it difficult for snakes to maintain a sustained physical effort for any length of time.

The gullet is very long, being over one-third the total length of the snake, and terminates in an elongated stomach. Both gullet and stomach are capable of considerable distension, thereby permitting large meals to be accommodated whole.

The liver is well developed and is quite a large organ some 13 cm in length and situated just behind the heart. The kidneys lie posterior to the testes of the male and the ovaries of the female. There is no urinary bladder, but both gall-bladder and pancreas are present.

Snakes do not possess an external ear or eardrums and cannot, therefore, hear sounds conducted by the air as can most animals. They do, however, possess an inner ear which is connected to the lower jaw bone, and it seems probable that they can hear sounds conducted through the ground, or whatever surface they happen to be lying on. They are certainly very sensitive to ground vibrations and can detect the lightest footfall many yards distant.

To compensate for the lack of eyelids the snake's eye is protected by a transparent scale, called the brille, which is part of the outer skin, known as the stratum corneum, which the snake sheds periodically. For a creature that spends its life in close contact with the ground this form of eye covering is, perhaps, an advantage, as particles of grit are prevented from entering the eye itself.

Instead of a single penis the snake has two, which are joined together at the base. This twin organ is known as a hemipenis, and its position is revealed by a swelling on either side of the spine at the base of the tail, more noticeable in the adder than in either the grass snake or the smooth snake. The two penes are used independently of each other, only one being inserted at a time into the body of the female. When erected the penis of the snake is turned inside out, revealing a surface covered with short soft spines. The spines help to prevent the penis from being dislodged from the female's body until an emission of sperm has taken place. Withdrawal is achieved by a special retractor muscle which turns the penis outside in so that the spiny surface does not injure the female. If disturbed during the act of copulation by an intruder the male snake is often unable to effect withdrawal immediately, due to the swollen state of the penis when erected and the spines holding it in position. I have on several occasions seen a male grass snake being dragged along by the female as they tried to escape in different directions on being disturbed whilst mating.

Tongue

While a snake cannot sting with its tongue, as is widely believed, it can and does use it in a way that is much more remarkable. The snake can, in fact, do with its tongue what we, and most other creatures, can only perform with the nose; that is, identify scents and smells. The snake's bifid tongue does not, however, replace the nose as an organ of smell, but simply augments it.

The rapid flickering of the tongue in and out of the mouth may appear to be nothing more than a nervous habit on the part of the snake, but in actual fact the tongue, when protruded and oscillated in this manner, is gathering scent particles from the air, and also from objects that it touches. When the tongue is withdrawn into the mouth the twin tips are brought into contact with two small ducts which open on to the palate. These ducts are connected to an organ, known as Jacobsen's Organ, which in turn is connected, like the nasal cavity, by nerves to the olfactory organ, or sense of smell.

A snake when in familiar surroundings and completely at ease—for example, when basking in the sun—makes very little use of the tongue. But the moment that it becomes aware, by vibrations transmitted through the ground, of impending danger, such as the approach of a large animal that might either accidentally trample on it or deliberately attack, the tongue is protruded and oscillated rapidly in an attempt to identify the intruder by its scent. The tongue is more frequently employed in this way when the intruder is beyond the range of vision, which in snakes is somewhat limited. The ground vibrations have, in the first instance, alerted the snake that some creature or other is approaching, and the snake relies upon its tongue to confirm the nature and proximity of a possible aggressor. When the air is very still and calm the snake may experience difficulty in picking up the scent of an intruder, and on such occasions the more usual rapid oscillations of the tongue are replaced by a slower and more deliberate action. The tongue is extended further and held motionless, curved back above the head, or down beneath the lower jaw, with only the finely tapered twin tips vibrating.

Use of the tongue, in the two different methods described, is also resorted to by the snake when hunting for food. When hunted by a snake many animals freeze into immobility in the hope of escaping detection. Because of its poor eyesight the snake quickly loses sight of its intended prey the moment it stops moving. When this happens the snake may also stop and wait for the creature it is pursuing to move again and thereby reveal itself. Frequently, however, the hunted animal sits tight, as often as not through fear, and the snake will begin to search about in the hope of locating it. This is when the tongue is brought into play, and all the while the snake searches the threadlike tongue is constantly flickering in and out through its closed jaws.

The adder relies to a considerable extent on its tongue to help it find prey at which it has struck and allowed to escape. When its stricken prey makes off, the adder does not follow in hot pursuit, but will wait for a few minutes, supremely confident in the lethal power of the venom it has injected to bring its victim to a permanent halt not too far away. Having waited a while, it will then trail the bitten creature unerringly by scent to its last hiding-place. Use is made of both nose and tongue on these occasions, and the adder will stop and press its nose to any spot of blood that its victim may have shed, while here and there it will pause to touch delicately with its tongue the herbage against which the bitten creature has brushed in its hurried flight. Every now and then it stops and waves its tongue slowly in the air, as it endeavours to locate by airborne scent particles the whereabouts of the dead or dying creature, which it always succeeds in doing, although scenting it out may take some minutes.

Locating prey and possible enemies is not the only use to which the tongue is put. During the mating season flickering of the tongue by the male along the female's back forms part of the courtship procedure. The purpose of this tender caress is not what it may seem though, for the male is in fact trying to discover by the odour given off by the female's skin whether or not she is ready to mate.

When withdrawn into the mouth, the snake's tongue is enclosed within a fleshy sheath which protects this highly sensitive organ from damage when large and bulky prey is being swallowed. Normally the tongue is only protruded with the mouth closed. There are exceptions, however, and the grass snake when feigning death usually allows the tongue to hang loosely from its gaping jaws.

When basking undisturbed a snake will sometimes indulge in a habit which I can best describe as tongue stretching. The tongue is extended to almost its full length and held either straight out or slightly curved, often with the bifid tip closed. There is no vibrating of the tips, and after being held almost motionless in this position for a few seconds the tongue is withdrawn; this procedure is sometimes repeated two or three times.

The adder performs this operation much more frequently than either the smooth snake or the grass snake, and its purpose seems to be to stretch or exercise the tongue and not the collecting of scent particles. It is most frequently seen to occur following a period of inactivity, and often before or after a bout of yawning, when the snake repeatedly opens its jaws very wide.

Of the three British snakes the grass snake makes more use of the tongue than either of the other two. When on the move the grass snake's tongue flickers constantly in and out, whereas the adder often moves about without protruding its tongue at all, except of course when hunting prey. The smooth snake uses its tongue even less, bringing it into play chiefly when hunting for food.

When in a defensive attitude the adder only occasionally protrudes the tongue, whereas the grass snake when alarmed and facing an intruder flickers its tongue continually in and out. The more nervous and alarmed it becomes the more frequently its tongue appears. If the intruder remains motionless a short distance away the rapid flickering of the grass snake's tongue often changes to a slow, waving motion, with the tongue more or less fully extended. The adder sometimes indulges in this slow waving of the tongue when defending itself, but not to the same extent as the grass snake. Sometimes the grass snake holds its tongue out stiffly, often with the fork closed, whilst at the same time emitting its warning hiss. This latter use of the tongue seems to be a threatening action, with the snake trying to convince its aggressor that its tongue is some kind of sting. I have never known either adder or smooth snake to employ the tongue in this manner, and in the case of the smooth snake the tongue is seldom exposed when danger threatens.

Locomotion

The snake appears to be in no way handicapped by its lack of limbs, and in its natural environment it can move at a speed that is equal, and sometimes faster than, many creatures with limbs. To say that a snake 'crawls' is to do it an injustice, such a term more aptly describing the locomotion of snails, worms, and similar creatures. Nor does the word 'glide' accurately describe a snake on the move. A snake when moving swiftly appears to flow in an effortless manner over the uneven terrain of its natural surroundings.

The snake has more than one method of locomotion, and can adapt its manner of progression to suit prevailing circumstances. When moving slowly it can advance over the ground with its body travelling in an almost straight line. If one watches closely one can see a ripling movement of the underside of the body, rather like that observed in a moving caterpillar, which gives the impression that the snake is walking on its ribs. In actual fact the ribs scarcely move at all, the rippling action being caused by the muscles attached to the ribs. In other words, the snake is walking, so to speak, on its muscles. I have frequently watched a captive snake employing this caterpillar action to climb the vertical sides of its cage. Although there may be nothing for the snake to hitch its body on to, it manages to travel in a completely straight line for some distance – as high as 35 cm in the case of an adder 60 cm in length – before overbalancing. The snake appears to be pushing against the floor of the cage to reach up the smooth walls of the cage, but closer observation reveals a pronounced caterpillar action being used by the snake to 'walk' up the vertical surface. This action can be observed even more clearly by watching a snake trying to climb a vertical sheet of glass.

When disturbed in its natural surroundings by an intruder, and when pursuing prey, a snake can move at a fast pace over a short distance with considerable ease. It appears not to use any particular style of locomotion for travelling at high speed under these circumstances, but simply flows among and over the vegetation quite effortlessly. On bare earth or close-cropped grass the snake cannot move anywhere near so fast or so easily, and when hurrying to escape it throws its body into a series of wide lateral loops in an attempt to obtain a better purchase. But this method of progression takes a lot of physical effort, and after travelling only a few feet the snake begins to tire and is forced to a more leisurely pace. The adder, because of its stouter build, is more rapidly exhausted by this form of locomotion than either grass snake or smooth snake. The grass snake is by far the most agile of the three British snakes, and can travel the fastest by this method. It also throws its body into much wider side to side loops than the other two, and I have known some grass snakes that could travel with a sidewinding action very similar to that of desert-dwelling snakes such as the Mexican rattlesnake, commonly known as a sidewinder. When using this method of locomotion the grass snake does not travel directly forward, but at an oblique angle to the direction its head is pointing. One gets the impression of a side to side rolling, with only the bottom of each of the deep lateral curves actually making contact with the ground.

The grass snake also employs a lateral movement of the body when swimming. Neither smooth snake nor adder are in the habit of swimming, although both have been observed to do so on occasion. For myself, I have never actually seen a smooth snake enter water, although I have known the adder to do so.

Besides being able to swim well the grass snake can also

climb, and will occasionally ascend into low shrubs to bask in the sun. I have seen a grass snake basking about 120 cm from the ground in the branches of a bush, the snake having climbed there to catch the rays of the rising sun which had not yet reached the ground below. The smooth snake is also a good climber and can climb a tree with coarse bark quite easily, a feat that is beyond both grass snake and adder. The smooth snake has the advantage of possessing the strongly muscular body common to all constricting snakes, which enables it to cling tightly, with any part of its length, to every small irregularity of the tree bark. In effect the whole of the smooth snake's body is prehensile, whereas the grass snake and the adder possess only prehensile tails. It is, however, doubtful whether the smooth snake does much climbing in its natural habitat, for feeding as it does on ground-dwelling animals it has no cause to become arboreal.

Hissing

The hissing sound made by most snakes is produced by inflating the lung with air and then expelling it forcibly through the glottis. Some snakes, the vipers for instance, hiss with both inhalation and exhalation, by sucking air rapidly through the glottis and expelling it with equal speed.

Yawning

After swallowing a large meal most snakes will indulge in wide yawns for the purpose of re-settling the jawbones – which are not connected posteriorly – back into their proper place. Those species of venomous snake, which, like the adder, have folding fangs, also yawn for the additional purpose of getting the poison fangs correctly positioned in the mouth. A snake when about to slough will often yawn persistently in an effort to break the old skin away from around the jaws.

A snake will also yawn while basking, even though it may not have taken food for several days, and when there is no possibility of it sloughing. By watching basking snakes I have found that a yawn is invariably indulged in when a snake has become thoroughly warmed by the sun, which may take between half an hour and one hour. Usually the yawn occurs when the snake changes its position, but not always, and I have known a snake to move several times while basking without making any attempt to yawn. In fact, when trying to photograph a snake yawning I have sometimes had to wait as long as three hours.

Of the three British snakes the adder yawns the most frequently when basking, and it invariably opens its jaws much wider than does the grass snake or the smooth snake. The jaws may be opened only once before the snake settles down again to continue basking, but as often as not several yawns are indulged in, particularly by the adder.

When about to yawn a snake usually lifts its head slightly from where it happens to be resting, which is either on its own body or on the ground. Having raised its head the snake usually pauses for a moment before opening its jaws, and occasionally there is a period of tongue stretching just prior to the yawn.

When a yawn takes place the jaws may be only partly opened at first. The jaws are then closed and two or three half-hearted attempts at a yawn may be made before the jaws are opened really wide. In the case of the adder the jaws are usually opened to their fullest extent with the poison fangs erected and pointing forwards. When indulging in a really wide yawn the adder's jaws are at an angle of 170 degrees, whereas the yawns of both grass snake and smooth snake seldom extend the jaws beyond 140 degrees, and often much less.

As the jaws close the snake sometimes reacts involuntarily, and the forepart of the body – occasionally the whole of the body – jerks violently sideways. This reaction can best be likened to the way in which we ourselves sometimes 'jump' when we are on the point of falling asleep.

After a single yawn, or perhaps a bout of two or three yawns, the snake usually continues basking, and normally

1. The grass snake makes more use of the tongue than either the smooth snake or adder, and usually extends it further when trying to identify the nature of an approaching intruder.

When on the move and when excited or agitated snakes flicker the tongue rapidly in and out of closed jaws.

2. The snake uses its tongue to gather scent particles from the air, and in this way it can detect the presence of its prey or the approach of an enemy.

3. When attacked or molested the grass snake will sometimes feign death by rolling on to its back and lying inert with its tongue hanging limply from its open jaws.

4. The adder always points its head in the direction from which it expects to be attacked. In this instance the adder's attention is directed towards an anticipated attack from overhead.

5. The adder delivers its poisonous bite by means of a stabbing action, the poison fangs being thrust forcibly into the flesh of either prey or aggressor. The fine tips of the twin poison fangs can be clearly seen protruding from their protective sheaths. Although taken with an exposure of only 1/3000th of a second even this photograph shows a slight blurring of the rapidly moving head and forepart of the body.

6. Male adder striking. The glove finger conceals the trigger mechanism for the camera.

Opposite
8. Courtship and mating of the adder (pls 8–12). At first the male makes no vigorous attempts to mate with the female, but simply basks with her in the sun. The male frequently lies with his head on top or beside that of the female.
9. The male in possession of a female will chase off other males that venture near. If unable to escape his pursuer the interloper will turn and face him, and the two of them will then fence with each other, the fore-part of their bodies raised above the ground.

7. Male adder yawning. The adder seen here has just used its poison fangs for defensive purposes, and now indulges in a wide yawn for the purpose of re-settling them into the correct recumbent position.

10. Having basked beside the female for some time, possibly two or three days, the male's urge to mate becomes stronger, stimulated by an odour given off by the female's skin. The male's eagerness to mate is indicated by the lashing of his tail from side to side, as he moves over and around the female, constantly playing his tongue over her body.

11. During courtship male and female will sometimes race wildly about together. The pace is incredibly fast and yet the two snakes move almost as one.

12. Try as she may the female is unable to elude the pursuing male, and the moment she tires and comes to a halt he immediately attempts to mate with her. He holds her tail firmly with his so that he can bring his vent into contact with hers. His behaviour is one of great agitation and excitement, and all the while he is trying to mate with the female his tongue flickers incessantly over her body, and his lower jaw taps against her back.

13. The male adder has mounted the back of the female slow-worm in the same way as it would when attempting to mate with a female adder.

14. Some time in July the female grass snake burrows into either loose soil or rotting compost to lay her eggs.

15. Grass snake emerging from egg. To enable it to escape from the egg the infant snake is provided with a special egg-tooth, which can be seen in this photograph as a small chisel-edged projection just below the tip of the nose. A few hours after the snake has left the egg the egg-tooth is shed.

will not yawn again for another hour or more. The habit of yawning is not confined to adult snakes; I have seen the young of all three British snakes yawn while basking.

As yawning in snakes occurs most often following a period of inactivity it would seem to indicate a need for extra oxygen, as with ourselves when we awaken from sleep. Many kinds of animals yawn after a period of rest. Dogs and cats, for instance, often indulge in yawns on waking from a nap, and they, like ourselves, will also yawn when bored. It is unlikely, however, that snakes suffer from boredom, being of much lower intelligence, and their habit of yawning while basking probably stems from a desire to stretch cramped jaw muscles, plus a need for more oxygen. The fact that snakes almost invariably change their basking position just prior to, or immediately after, yawning would seem to support this theory.

Life-span and Growth

The maximum life-span of British snakes has never been ascertained, but from the information available on the growth of snakes it can be roughly calculated as being somewhere in the region of twenty years, while the average life-span is more likely to be in the neighbourhood of fifteen years, where violent death does not intervene.

One of the earliest observations on the growth of the grass snake was made by Gadow who kept a specimen in captivity for a period of nine years, during which time it grew from 89 cm to 106 cm, which gives an average rate of growth of 2 cm a year.

Carl Edelstam of Stockholm more recently found by marking an almost fully grown male grass snake, which he released and then recaptured in four successive years, that it had grown at an average rate of 2 cm per year.

The female grass snake grows more quickly, and to reach sexual maturity and a length of 65 cm takes it four years. From this age and length, growth continues at the rate of approximately 3 cm each year until a length of 80 cm is reached, by which time the snake would be nine years of age. From this point onwards growth becomes slower, averaging 2 cm per year. Therefore, to reach a length of 106 cm would take a female grass snake approximately twenty-two years. However, in this country the grass snake seldom exceeds 91 cm in length, which would give an average life-span of about fifteen years.

A similar life-span for the adder is arrived at when its rate of growth is estimated from the known facts on the subject. Volsoe's observations on the growth of adders show that the female is sexually mature when four years old and measures 47 cm in length. From this stage growth continues at an average rate of 2 cm per year until a length of 55 cm is attained. At this length the adder would be eight years old. From then onwards growth is slowed to 15 mm each year, at which rate it would take an adder a total of fifteen years to reach the average length for a fully-grown female, which is 66 cm.

A snake's growth is governed chiefly by the temperature of its environment. At high temperatures its metabolism is more rapid, and its intake of food is therefore greater than at low temperatures. Consequently a snake will grow more during a warm, prolonged summer than during a short, cool one. This explains why, as records have shown, a snake exhibits varying rates of growth over successive years. During hibernation no growth at all takes place, which means that a prolonged winter, or a succession of prolonged winters, could also have some effect on a snake's rate of growth over a given number of years.

The rate of growth in individuals also varies, and therefore it is not possible to state with any degree of accuracy by what amount a snake will increase its length each year. Growth is rapid in the early stages of life, becoming progressively slower approaching maturity, and decreases markedly once maturity has been reached.

As a snake's growth lengthwise decreases with age so its girth tends to become larger. An adder 62 cm in length is usually about twice as big in girth, and possesses a much broader head than a specimen 55 cm in length. This applies equally to the grass snake, and a specimen 90 cm in length will appear a giant beside a specimen 76 cm in length, although the difference in their lengths is only 14 cm.

Intelligence

In the main snake behaviour conforms to a regular pattern and is largely instinctive; most of the snake's natural functions such as feeding, reproduction, and hibernation, having been programmed, as it were, into the reptile's brain prior to birth. One finds little deviation from the general pattern of behaviour, and this suggests a fairly low level of intelligence. Nevertheless, snakes do seem to be capable of learning from experience, as has been illustrated to me by the behaviour of captive snakes.

When wishing to keep specimens under close observation I have sometimes housed them indoors in a large cage with a glass front, and a trapdoor in the top through which food and water can be introduced. On occasion the cage has contained three or four adult grass snakes which, when first made captive, invariably dive for cover whenever the trapdoor is opened. To ensure that all the inmates of the cage receive an adequate share of food it is necessary to supply this at more frequent intervals than would be the case if only one snake occupied the cage. Consequently, the snakes quickly learn that the opening of the trapdoor means that food will be placed inside the cage, and they cease to panic when this occurs. If the trapdoor is opened while the snakes are hidden beneath the large pieces of bark provided for them, they will emerge a minute or two later to look for the food that has been introduced into the cage. This becomes such a habit with them that eventually it is only necessary to open the trapdoor, without putting food into the cage, to draw them from their hiding-places.

They also learn that food is invariably placed in their water bowl, and on emerging from beneath the bark they make straight for the bowl. If they find nothing in the bowl, they will search around it for a while, and then settle down to bask. As the water bowl is always in their cage there is no question of their going to it just for a drink after the trapdoor has been opened and closed. As a rule the snakes will make for the bowl shortly after the trapdoor has been closed and gaze into the water. If they find nothing there they simply move away again, although sometimes a snake will immerse its head in the water and nose around beneath the surface in the hope of finding something edible. On finding nothing the snake will lift its head out of the water and, without pausing for a drink, move off to another part of the cage.

If two snakes headed for the water bowl at the same time one would invariably hold back until the other had investigated the water for signs of food, and would not as a rule approach it until the first snake had moved away.

I have found too that the majority of adders soon learn that striking at inanimate objects thrust towards them is completely futile, and they eventually refuse to strike at such things. Many adders when first captured strike wildly in all directions, even at the empty air, but after only a few days in captivity an adder, if regularly picked up and handled, soon ceases to panic and will allow itself to be touched if gently treated and no sudden movements are made near to it. Indeed I have found that many adders are more easily tamed than are some grass snakes, which often continue to struggle to escape when first picked up, although they may have been handled on numerous occasions before. Smooth snakes are also easily tamed, but whether or not this indicates a higher degree of intelligence compared with a snake that remains intractable is difficult to decide.

I now make a point of handling snakes as little as possible after they have been taken in to captivity, and I am writing now of my earlier experiences with snakes.

The fact that many snakes seem unable to recognise their normal prey unless it shows signs of movement, and after they have examined it at close quarters, would seem to indicate a very low level of intelligence. It denotes that in most cases feeding is an automatic response to certain stimuli, and that when such stimuli are lacking feeding does not take place, no matter how great the snake's hunger. This would seem to put the snake's intelligence on a level with much lower forms of life.

Enemies

Man is the chief enemy of the snakes of this country, because he destroys their natural habitats by agricultural and urban development. He is also inclined to kill them on sight due to his apparently inherent dislike or distrust of snakes generally.

Mammals that prey occasionally on snakes include the hedgehog, badger, and fox. Many birds of prey are known to kill and eat them, but there is no animal indigenous to Britain that feeds habitually or exclusively on snakes.

The infant snake is, perhaps, more vulnerable to being preyed upon by other animals, as being so small and worm-like it could easily be devoured by non-predatory birds and quite small rodents. Even the common toad has been known to eat very small snakes, possibly mistaking them for worms.

An additional hazard to the snake's existence is created by man in the form of fire. Every year fires on commons and heathlands result from the careless throwing away of lighted cigarette ends and empty bottles. It is estimated that the majority of these fires are caused by empty bottles left lying about which focus the sun's rays on to the surrounding herbage. The danger of fire is particularly great during the spring when dead grass and bracken are often as dry as tinder and therefore easily ignited.

Eyewitnesses to the large scale forest fire which occurred in Thetford, Norfolk, some years ago reported seeing hundreds of snakes wriggling away as fast as they could go from the fire. I have witnessed more than once the speed with which fire can spread across a common. I can recall on one occasion walking slowly across Arbrook Common in Surrey with flames from burning heather and dry grass following close at my heels. There was only a light wind blowing at the time. Had the wind been any stronger the fire would have spread much more rapidly, and at a speed too fast for a snake to be able to escape from it. The only way a snake, or any other small animal for that matter, could save itself from a fire of that sort would be to go underground and remain there until the fire had burnt itself out, which would not take long. But that a snake in that kind of situation would have the sense to stay below ground is doubtful. The smoke from the fire might drive it into the open where it would quickly perish. Even if most of the snakes within the area of the fire were fortunate enough to survive, their natural habitat would have been destroyed and they would be forced to find a new environment.

Fire can devastate the whole ecological structure of a large stretch of commonland, as it would destroy not only the vegetation but also most of the insect life. As the prey of snakes consists largely of insect-eating reptiles and amphibians, which would be forced to leave the area through lack of food, the snakes too would have to depart. In any case it is unlikely that many animals would remain in an area denuded of all form of cover.

In time of course the burnt area would again be covered with vegetation, although just how long this would take depends to some extent on the time of year when the fire occurred. If the fire were to take place in the early spring it is likely that new vegetation would have grown up by the end of the summer, whereas if the fire occurred in late summer new vegetation would probably not appear until the following year. It would take a little longer, possibly a year or more, before the wild life driven out by the fire would return.

Parasites

Lacking both fur and feathers the snake does not harbour as many external parasites as do mammals and birds. Almost the only external parasite of the snake is a small globular-shaped red mite about the size of a pin-head, which sucks the blood of the snake. It is a carrier of the bacterium *Proteus hydrophilus*, which causes an hemorrhagic septicaemia that is fatal to the snake. Fortunately for the snake, these mites are easily rubbed off as it moves about amongst dense herbage, and they also get washed off when the snake enters water. Possibly for this reason snakes living in their natural environment seldom carry these mites. I have never found them to be present on any of the British species of snake that I have captured, although I have kept a careful watch for them.

Captive snakes are more prone to snake mite than those living in the wild, and most probably because the captive snake is unable to rid itself of them so easily. In the early stages of infestation when there are only a few mites present they are not easily detected, since when the snake is picked up and handled they usually disappear quickly beneath the reptile's scales. They multiply rapidly, however, and when a snake is badly infested with them they rub off on the hands the moment the snake is picked up. When a snake is carrying only a few of these mites they are more easily detected when the snake is basking, as they can then sometimes be seen moving slowly along the snake's body, pausing here and there to drink its blood. That a snake is in fact the host of snake mites is eventually indicated by the development of abscesses beneath the scales, which show themselves as small hard swellings that gradually increase in size. By the time this symptom has developed, however, it is usually too late to prevent the snake from dying.

If the presence of snake mites can be detected before abscesses start to form, the snake's life can often be saved by bathing it liberally with medicinal paraffin, afterwards placing it in a clean cloth bag. The snake's cage should then be thoroughly scrubbed with hot soda water and allowed to dry before the snake is returned to it.

Diseases

A fairly common ailment to which captive snakes are prone is that known as canker, which first affects the mouth and throat and finally the skeletal structure of the body. The first symptom is the appearance of a white fleshy growth on the membranes lining the mouth which increases steadily until the mouth is filled and the snake is unable to close it properly.

When affected by this complaint the snake shows no inclination to feed, and becomes increasingly lethargic as the canker spreads. Sometimes the throat becomes enlarged, and often the eyes are affected, fluid collecting between the brille and the eye itself, making the eye appear to stand out from the head. A clicking sound emanating from the body of the snake when it moves indicates that the disease has affected the bones.

T. H. Gillespie states in his book, *The Way of a Serpent*, that canker in snakes can be cured if caught in the early stages. Treatment consists of bathing the mouth with a mild antiseptic fluid, such as diluted peroxide, and afterwards dusting it with a dressing powder.

As with septicaemia, canker of the mouth is a bacterial disease, although exactly how a snake contracts it is not known for certain. It has been suggested that an injury to the mucous membranes of the mouth may permit entry of bacteria. While the cause is uncertain it seems likely that the condition is infectious, as snakes subsequently kept in a cage that has housed a snake afflicted with this malady have been known to contract it. I have, however, known healthy snakes to share the same cage as a snake affected with canker without their contracting the disease.

Snakes have also been known to contract pneumonia, although this most usually occurs in imported snakes which have come from a warm climate and have been kept in captivity at too low a temperature.

In my experience British snakes living in their natural habitat are never subject to any of the diseases mentioned, nor do they develop them in captivity. Septicaemia, mouth canker, and pneumonia seem to be confined mainly to snakes imported from other countries and kept here in captivity.

Malformations

A condition which I have occasionally found in grass snakes and adders living wild in this country is blindness or malformation of one eye. I have caught grass snakes in which one eye has been faintly obscured by a whitish mist, and slightly sunken into the head. This semi-opaqueness of the eye has not been due to the brille or eye scale not having been removed when the snake has sloughed, as sometimes occurs. Usually the eye itself has been malformed in some way. I have also seen a female adder in which the pupil of the left eye was malformed, being horizontally elliptic in shape, and showing no response to light by either contracting or expanding. The eye itself also appeared incapable of the small amount of movement that a normal snake's eye possesses, and was fixed in a downward looking position. It is possible that the eye had been injured in some way, although there was no sign of damage either to the eye or to the scales or plates surrounding it, and I felt inclined to believe that the adder had been born with a defective eye.

I have never come across any other kind of malformation in British snakes, although Dr Leighton mentions in *British Serpents* a case which appears to be authentic. It concerns a two-headed adder, about 30 cm or so in length, that was found at St Martin in Cornwall in 1853. The adder was sent to the London Zoo where it arrived alive, but died shortly afterwards. Judging by its length it must have been about one year old.

Two-headed snakes of different species have been recorded in other countries, but they are very rare, and seldom survive for very long after birth.

CHAPTER THREE

Courtship and mating

Mating season – basking period – male ardour and female reluctance – rivalry among males – change of partners

THE mating season for all three British snakes usually begins in early April and extends over a period of about six weeks. Mating does not take place as soon as hibernation has ended, for snakes require a period of warmth to revitalise them before mating is possible. An early spring will bring snakes out of hibernation sooner and, as a result, mating will start a little earlier than when emergence from hibernation is delayed by low temperatures.

The adder is usually the first to appear, and it has been seen out basking as early as the end of February in some parts of the country. The earliest date I have found adders out basking in the south-east of England is 3 March, and on that occasion the temperature reached a maximum of 14·5° C in the shade. In my experience a shade temperature of at least 13° C is needed to tempt the adder from its winter hiding-place, while both grass snake and smooth snake do not normally make an appearance until the shade temperature reaches 15° C, which seldom occurs before the middle of March. However, as temperatures tend to fluctuate widely throughout March, hibernation does not normally finish until about the end of the month.

For the first week or two following emergence from hibernation snakes spend the sunny hours basking, providing the temperature is high enough, while on sunless days they rarely make an appearance. By about the second week of April the temperature has usually risen enough to arouse in the male a desire to mate, and he begins searching for a female. Where several snakes of both sexes have spent the winter in communal hibernation the male does not have far to look for a mate. But even the male that has hibernated alone will have little difficulty in finding a female, for at this time of year she gives off a scent by which he can trail her.

This scent or odour is produced by the anal gland situated at the base of the tail. In the grass snake this scent is particularly strong, and in the spring I have sometimes been able to detect the presence of a female nearby by means of it. In the adder and smooth snake the scent from the anal gland is less pungent, but can sometimes be detected. I can recall capturing a female adder one spring which opened the vent in the manner of a grass snake. There was no discharge from the vent, as occurs with the grass snake on capture, but an odour, similar to that of the grass snake but much fainter, was distinctly noticeable.

During the mating season the female will often move about with the vent open. This is done apparently to release the anal gland secretion which becomes smeared on vegetation that the female brushes against, and enables the male to locate her. She will also at times keep the vent open for quite long periods while basking, seemingly for the purpose of releasing scent to attract the male.

Before mating takes place there is always a period of courtship which may occupy several hours or even days. Young males are invariably more ardent than older ones and will sometimes attempt to mate with a female after only a brief courtship. From my own observations I have found that the male has to be sufficiently warmed by the sun before he will begin courtship, and as temperatures tend to vary considerably from day to day at this time of the year, he may spend several days basking beside the female before the temperature rises high enough – around 18° C in the shade – to arouse in him the urge to court her. Even on a warm day it is usual for the male to bask close beside the female for about an hour or more before an involuntary twitching of his tail indicates that he is ready to begin courtship.

Having lain quietly up until this moment the male now begins to move around and over the female, his head jerking back and forth in a seemingly uncontrollable manner. As his excitement increases this jerking movement extends to his body, while at the same time he plays his tongue incessantly over the female's body. This tongue movement is more rapid than when the organ is used on other occasions, and only the bifid tip is protruded, although the grass snake sometimes extends the tongue more fully and plays it slowly over the female he is courting. As the male's head jerks back and forth it also appears to see-saw up and down, so that his chin is constantly tapping against the female as he plays his tongue over her.

The obvious excitement of the male is not matched by the female, and she continues to bask, seemingly indifferent to the amorous attentions of her suitor. This in no way deters the male who becomes more and more excited, his tail lashing from side to side as he tries to push his head in between her coils in an attempt to make her unwind, so that he can bring the full length of his body into contact with hers. His feverish desire to mate is stimulated by the scent given off by the female's skin, which the male is able to detect with his tongue, and accounts for the continual use of the tongue by the male during courtship. It is the scent exuded from the female's skin that actually arouses the male's urge to copulate, whereas that given off by the anal gland serves mainly to bring male and female together. The scent given off by the skin is stronger at certain times during the mating season than at others, and this explains why one female may attract the attention of one or more males, whilst another female is completely ignored by them.

If, after an hour or two of courtship, the male gets no response he may relax his efforts and lie quietly basking for a while on top of or close beside the female. Often, however, he becomes so frustrated in his unsuccessful attempts to arouse the female's interest that he will go off by himself, moving swiftly about nearby in a highly agitated manner. Having worked off his excitement in this way, he begins to look around for the female once more. But he appears to have forgotten where he left her, and he searches about in a feverish manner, like that adopted by a snake when hunting for prey it has seen and then lost sight of. When he at last re-discovers the whereabouts of the object of his desires he once again becomes convulsed by involuntary jerkings of head and body, and courtship continues as before.

While the courtship behaviour of the male is much the same in all three British species of snake, the reaction of the male in possession of a female to the approach of another male while courtship is in progress varies. Rivalry among male grass snakes for possession of a female has never been observed, whereas rivalry frequently occurs between male adders and also between male smooth snakes.

When an unattached male adder approaches a courting pair the male in possession will give chase and the two males will race through grass and heather at an incredible speed. Indeed, anyone who has not observed adders during the mating season would probably find it difficult to believe that this normally slow-moving reptile could attain anywhere near such speed. The male being pursued will often double back quickly in an attempt to evade his pursuer. Sometimes he succeeds, but more often the pursuing male turns just as quickly, whereupon the interloper, finding that his every move to get past his pursuer is intercepted, will adopt an aggressive attitude by rearing up from the ground. The pursuing male takes up a similar stance, and the two of them will fence with each other, ducking and weaving with the forepart of the body raised from the ground. These brief encounters – they last only a few seconds – may take place two or three times during the course of the pursuit, and are invariably ended by the interloper attempting to escape. Eventually he succeeds, and the male that was in possession returns to his mate.

I have never known the intruding male to come off best in these encounters. It is not, in my opinion, a case of right triumphing, but simply that the male in possession, having become stimulated to fever pitch by his frustrated attempts to mate, is the more excited and aggressive of the two males, and is frantically determined not to lose his chance to mate.

At no time during these displays of rivalry does the adder attempt to make use of its poison fangs, and in fact no injuries are ever inflicted. Male smooth snakes are less reticent in this respect, and rivals disputing possession of a female will bite savagely as they writhe entwined about each other on the ground.

Old females sometimes show a complete lack of responsiveness to the courting male, and stubbornly refuse to uncoil so that mating can take place. The younger females are more easily disturbed from their basking positions, but seem no more eager to mate, and a young female adder will often try to escape the persistent attentions of the male by dashing off at high speed. A wild chase through the surrounding herbage, similar to that which occurs between the males, then ensues, the chief difference being that the female does not turn upon her pursuer and fence with him. The speed at which they move is quite astonishing, and throughout this wild chase their two bodies seem to move as one, every sudden turn and manoeuvre being carried out in complete unison. The two snakes appear to flow over and among the heather and grass, without any laboured side to side undulations of the body. The female seems intent on avoiding the male's amorous advances, but he is not easily shaken off, and when she eventually tires and stops he is right beside her. Immediately he will endeavour to position himself so that he can hitch his tail around hers for the purpose of bringing both their vents into contact. If ready to mate the female will permit him to do this, but if not she will coil up tightly, tucking her tail in close among her coils. But the chase has excited the male still more and he redoubles his efforts to mate with her.

I have never witnessed male and female grass snakes or smooth snakes dashing about together in the manner described. The female of both these two species simply keeps slowly on the move, or coils up tight, when trying to avoid the attentions of the male, thus making it difficult for him to make sexual union.

The male snake that obtains no response from the female he is courting will often go in pursuit of any other female that comes near, although if she also resists his advances, or is not emitting from her skin the scent that excites him, he will invariably return to his first chosen.

When ready to mate the female will uncoil and lie extended, whereupon the male glides quickly alongside her. With their two bodies touching he rests his head upon her back, and begins to move back and forth in an attempt to find the position in which he can bring his vent into contact with hers. His head and body jerk convulsively all the while, and he at last achieves union by lying either on top of or beside her and encircling the base of her tail with a single coil of his. Alternatively, he may simply push the base of his tail in an inverted position beneath the female. Sexual union, once achieved, often lasts for several hours.

It is not unusual for a male to temporarily lose the desire to mate, after having mated once or perhaps twice with the same female, and for the female to be taken over by another male. Later the first snake may regain the mating urge and either seek out another female or try and return to the first one. When a male adder attempts to regain a former mate he is invariably chased away by the male now in possession. The tables have been turned on him, and from being the pursuer he is now the pursued.

Although the mating season lasts for approximately six weeks, courtship and mating does not continue between the same pairs throughout the whole of that time. Having mated once, or perhaps twice, male and female usually separate, although both may mate again later with a different partner. As the warm and sunny weather required to induce mating is unlikely to prevail throughout the mating season, snakes must seize the opportunity to mate whenever conditions are suitable. Consequently the majority of matings tend to take place when the first settled spell of sunny weather arrives, although matings may still continue to occur amongst other pairs that have not previously mated until about the middle of May.

I have recorded finding on 13 April 1952 three pairs of grass snakes in copulation, all within a few yards of one another, on Arbrook Common in Surrey. They were discovered among long grass in the middle of the common, and close to a narrow belt of birch trees. A little further away I also found an unattached male. The weather at the time was mostly sunny, and the shade temperature reached a maximum of 18° C. The following day I discovered another pair of grass snakes copulating on Coneyhurst Hill, in Surrey, while on 15 May of the same year I came upon another pair of grass snakes in sexual union on Arbrook Common.

Also recorded in my notebook is the finding, on 18 April 1954, of two pairs of grass snakes in copulation a few yards from each other on Arbrook Common. The snakes were found in a wooded part of the common where the trees were mainly of oak, well spaced to permit the entry of sunlight and the abundant growth of bracken and brambles. There was a north-easterly wind blowing that day, and the temperature reached a maximum of 14·5° C in the shade. However, the sky was almost cloudless, and in the shelter of the trees where the snakes were seen the temperature in the sun was probably in the region of 21° C, or perhaps even higher.

I took one of these two pairs into captivity so that I could make closer observations, and they coupled again on 1 May and also on 6 May, after which they made no further attempts to mate.

In the spring of 1968 I recorded an unusual incident involving a male adder that attempted to court a female slow-worm, although there were female adders available. The adder in question had been exploring its enclosure when it came upon the slow-worm lying coiled. Its interest was immediately aroused, and I fully expected it to make a meal of the slow-worm. Instead it began pressing its nose against the slow-worm and touching it with its tongue as it would do if courting a female adder. Then the adder partly opened its jaws and pressed them against the slow-worm's back as though about to devour it. But the adder made no attempt to bite the slow-worm, nor did it try to take hold of it in its jaws.

After a short interval the same thing was repeated, after which the adder began again to prod and poke the slow-worm with its nose as before. This continued for about ten minutes, during which time the slow-worm remained unmoving.

As there was no response whatsoever from the slow-worm the adder took itself off to a spot a few inches distant, where it settled down to watch the slow-worm. The slow-worm remained motionless for quite some time, and when at last it did begin to move the adder edged towards it as if about to strike. Such behaviour would have been normal, but what followed was certainly not. For as I watched in astonishment, the adder mounted the back of the moving slow-worm in exactly the same way as it would a female adder when attempting to mate with it. The slow-worm appeared not in the least alarmed and did not hurry itself, but moved slowly across the floor of the enclosure with the adder on top of it. Just then the adder appeared to catch a glimpse of its own reflection in the glass front of the enclosure, and leaving the slow-worm it rushed in a seemingly aggressive manner towards its own reflection, its head held well above the ground as though about to begin combat with what it obviously thought was a rival male adder. It soon realised its mistake and went off hurriedly after the slowly departing lizard.

When later the slow-worm was removed from the enclosure the adder began to follow a female of its own kind, with which it attempted to mate on two separate occasions. But when the female made off the male quickly lost interest and made no attempt to pursue her. Indeed, it did not appear to be stimulated to the same extent by females of its own kind as by the female slow-worm. When the slow-worm was again returned to the enclosure the adder renewed its courtship behaviour, poking and prodding at the slow-worm as it lay quietly basking.

From the adder's behaviour it seemed probable that the slow-worm was giving off a scent which aroused the sexual interest of the male adder and, feeling more inclined to mate than to feed, it did not recognise the slow-worm as its prey. It is also possible that the adder was not accustomed to feeding on slow-worms and simply mistook it for a female of its own species, although I should have thought that the scent of one was quite different from the other. I should perhaps add that the adder in question was made captive only a few days prior to the incident related taking place, and in early June when the mating season has usually ended.

I witnessed another strange incident, involving two adders this time, a few days prior to the one just related. I feel it is worth mentioning since it relates to a point of behaviour mentioned in the previous account.

On this occasion I was watching several captive adders basking together, among them being a very large female measuring 66 cm in length, and a young male of only 35 cm. Although the female was nearly twice the length of the young male, and several times bigger in girth, her size did not deter him from attempting to court her. At least it appeared to be a form of courtship, although the male could not have been sexually mature judging by its size. However, after the small male had lain basking beside the big female for several days he began touching her with his tongue. Eventually he climbed on to her back, which was so broad that he was able to travel back and forth along it as though it were a roadway. He continued touching her with his tongue in the manner of a courting male, and then, to my surprise, he opened his jaws and pressed them against the female's back. I could see that the young male's poison fangs were still folded, and that he was making no attempt to bite her, but simply pressing his open jaws gently against her body. This was repeated several times, and although I tried to photograph the incident the young male adder would always close his jaws just as I was on the point of making the exposure. Consequently I failed to get a picture of this extraordinary incident.

Although mating among British snakes normally occurs during April and May, matings have been known to take place during the summer, and also in the autumn. Combat between male adders has also been observed in the autumn.

I have never witnessed mating among British snakes outside the normal mating time, but I have often seen the jerking movements of the head, associated with mating, indulged in by captive snakes at various times during the summer and long after mating has finished. I have found that it most usually occurs when a male makes physical contact with a female, and is more likely to occur in male grass snakes outside the mating season than in adders, while smooth snakes seldom behave in this manner except during the mating season.

It would seem that this particular form of behaviour is inherent in the male from birth, and is not acquired with sexual maturity. For I have observed the very same head-jerking behaviour in young grass snakes only a month old when a number of them have been basking together.

CHAPTER FOUR

Reproduction

The laying and hatching of grass snake eggs – birth of adders and smooth snakes – infant snakes' awareness of danger

ACCORDING to their method of reproduction snakes are classed as being either oviparous or ovo-viviparous. The former lay eggs containing embryo snakes in an early stage of development which will not hatch for several weeks, while the latter retain their eggs within the body until they are on the point of hatching.

GRASS SNAKE

Of the three British species of snake the grass snake alone is oviparous, and about mid-June the female begins to search around for a suitable place in which to deposit her eggs. As moisture is essential to the development of the eggs after they have been laid she must choose a site that remains constantly damp, but which is not likely to become waterlogged. It must also be a place where the temperature is fairly constant, and for this reason manure heaps and piles of decaying vegetable matter are favoured as a depository for the eggs. Having found a suitable spot the female remains in its vicinity until the eggs are ready to be laid, which is usually some time in July, although egg-laying can take place at any time between the end of June and the middle of August. The actual time of egg-laying is largely controlled by the date when fertilisation of the eggs takes place. A mating in mid-April would result in the eggs being laid towards the end of June, whilst if mating did not occur until the end of May, egg-laying would be delayed until about the middle of August. The weather following fertilisation also determines, although to a lesser extent, the time of egg-laying, for development of the eggs within the female's body is controlled by the temperature of the environment. Sometimes a late mating and a cool summer combine to delay egg-laying beyond August, and I have recorded eggs being laid by a grass snake in mid-October. It is unlikely that eggs laid so late in the year would hatch the same year, and the possibility of their surviving the winter is very doubtful. However, the French naturalist Rollinat has recorded finding grass snake eggs on 14 November which contained young on the point of hatching.

The female grass snake knows instinctively the right conditions required for development of the eggs after they have been laid, and as suitable conditions are not everywhere available it is not unusual for a number of females to gather at one spot for the purpose of egg-laying. One such place existed some years ago on the edge of Ashtead Forest in Surrey, where, in the month of June, one could expect to see on a sunny morning as many as six or more female grass snakes basking within a few feet of one another. Doubtless there were many others out of sight, and all of them were attracted to this particular spot by a large amount of sawdust which covered an area of several square yards, and was three feet deep in places. The sawdust had resulted from wood-cutting operations some years previously, and it had lain where it was for so long that it was almost completely overgrown with bracken. Being in a state of decay it generated a warmth of its own, besides retaining the right degree of moisture to make it an ideal incubator for grass snake eggs.

Numerous similar instances where large numbers of grass snakes have collected at one spot for the purpose of egg-laying have been recorded in different parts of the country. Leighton mentions in his book, *British Serpents*, the finding of forty bundles of grass snake eggs in the wall of a house at Llanelly in Carmarthenshire.

Not all grass snakes utilise compost heaps and the like as a depository for their eggs, and many lay them in the loose soil around the roots of a bush, beneath rotting logs, or in the disused burrows of other animals.

When the eggs are ready to be laid the female grass snake burrows into whatever material she has chosen, and by moving round and round, while at the same time pushing outwards with her body, she forms a cavity to take them. Emergence of the eggs from the oviducts is fairly rapid at first, one following the other at intervals of about five minutes. The last few eggs invariably take longer, and I have known the time lapse between the emergence of one egg and the next to be as long as thirty-seven minutes towards the end of egg-laying. The time taken to lay all of the eggs varies, and depends to some extent on the temperature of the snake's surroundings. I have recorded one female grass snake 71 cm in length as having taken two hours to lay fifteen eggs, while another grass snake 76 cm in length took approximately five hours to lay nineteen eggs.

When they first emerge from the female's body the eggs are covered with a mucous secretion from the walls of the oviducts which dries rapidly on exposure to the air. The movement of the snake around the eggs as laying proceeds forms them into a cluster or irregular shaped bundle, and the mucous cements them together as it dries.

Normally all of the eggs are laid at the same time, but occasionally some of the eggs are retained and may not be laid until a few days later. Having laid her eggs the female may remain with them for a day or two, not for the purpose of protecting them, but simply because she is exhausted by her efforts. She eventually departs, leaving the eggs to their fate.

White in colour and elongated in shape, the eggs measure approximately 25 mm by 15 mm when first laid, and each egg is covered with a tough, leathery skin. The eggs in one clutch may be of uniform size, but this is not always the case. Quite often the eggs laid by one female vary in size, some being smaller and some larger than the average. The number of eggs laid varies according to the size of the grass snake, and the number of eggs laid by a female of average size, which is about 76 cm, is between twenty and twenty-five.

The largest number I have recorded being laid was thirty by a grass snake measuring 89 cm in length. A greater number of eggs in one clutch is not unusual, the most recorded from one grass snake being fifty-three.

Moisture is absorbed by the egg after it has been laid, causing it to increase in size, and those laid in very moist surroundings tend to become larger than those in dryer situations. I have found with grass snake eggs hatched in an incubator that those kept only slightly moist showed little increase in size from the time they were laid until they hatched, whereas those in almost waterlogged sand increased enormously, and some were found to measure as much as 35 to 40 mm in length when about to hatch. The size of the egg, however, bears no relation to the size of the embryo snake, and the snake that emerges from the large egg may be no bigger in size than the snake that hatches from a small egg.

Eggs which obtain very little moisture from their surroundings tend to shrivel a little as the time for hatching approaches, whereas those laid in very damp situations remain full and rounded until the moment of hatching. The skin of the well moistened egg seems to become thinner as the result of being stretched by the swelling of the egg. It also becomes semi-transparent, enabling the embryo snake within to be seen in silhouette when a light is placed behind the egg. It is also possible to see quite clearly the blood vessels with which the skin of the egg is richly endowed. The skin of the egg that has absorbed little moisture remains completely opaque.

All the eggs laid by one female will often hatch on the same day, but sometimes hatching of the eggs in one clutch extends over several days. There are usually one or two infertile eggs in a clutch, but as a rule most of the eggs laid hatch successfully. When the moment arrives for the infant snake to emerge it begins pushing and thrusting against the skin of the egg and these movements can be clearly observed. If the skin of the egg is very moist and thin the young snake can sometimes push its head straight through, but it is more usual for the young snake to slit the skin with its egg-tooth.

Often several slits are made as the young snake slides the tip of its nose back and forth along the inside of the egg walls in an effort to escape, and some of the slits may be almost as long as the egg itself .

The egg-tooth is situated below the tip of the nose and is so minute that it is difficult to see with the naked eye. When viewed from the side under magnification it appears to be sharply pointed, and shaped like a tiny upcurved thorn. But when viewed from above or below it will be seen to be flat and chisel-shaped, sometimes with a groove down the middle, and tapered to a fine cutting edge.

Having succeeded in piercing the egg wall the infant snake pushes its nose through the aperture it has made, and the tiny forked tongue that comes flickering out will be seen to have a white tip, unlike that of the adult snake which is entirely black. By picking up scent particles with the tongue the young snake can tell if there is danger lurking outside, and if it is satisfied that all is well it will then push its head out and take its first cautious peep at its surroundings. Having got that far it makes no attempt to emerge further, and may remain for several hours, up to twenty-four sometimes, just looking out. The reason for this is not that the young snake is reluctant to leave the comparative safety of the egg, but simply that it is still attached to the egg by its umbilical cord. By watching closely it is possible to see the periodic movements of the young snake's body within the egg as it endeavours to free itself from the connecting cord. While this struggle is in progress the infant snake's head remains unmoving where it protrudes from the aperture in the egg wall. In some cases the baby snake is able to free itself from the umbilical cord shortly after slitting the skin of the egg, and may then leave the egg almost immediately.

At the slightest hint of danger the baby snake withdraws at once into the egg, and ten to twenty minutes may pass before it will venture to poke its nose out again. When at last ready to leave the egg the tongue is once more brought into play, and then slowly and cautiously the young snake emerges and slithers away.

When first hatched the infant grass snake is much darker in colour than the adult, but after shedding its skin for the first time, within two to three weeks of hatching, the baby snake emerges in the colours of the fully-grown snake.

The average length of a newly hatched grass snake is 17·2 cm. The smallest that I have measured was only 15·2 cm in length, the largest 19·3 cm.

Adder and Smooth Snake

Both the adder and the smooth snake are ovo-viviparous, the eggs being retained within the body until the embryo snakes have developed to the stage where they are able to begin an existence independent of the parent. For several weeks prior to birth of the young the female spends much of her time basking in the sun. Being cold-blooded she relies upon external heat for the development of the young within her, and the more time she can spend basking the quicker their growth. This means that the young will be born slightly earlier than usual in a hot summer, and later in a cool one, although the chief factor controlling time of birth is the date when mating, or fertilisation of the eggs takes place.

There is little discernible enlargement of the female's body until she is within a fortnight of giving birth, when the latter two-thirds of her body begins to look noticeably distended. By this time the weight of the almost fully developed young she is carrying makes it difficult for her to move very quickly, and for this reason she becomes more secretive than usual in her behaviour, and takes care to bask close to, or in amongst, dense cover. She makes no special preparation for the arrival of her young, and gives birth to them under cover of heather or grass near to her hiding-place. Each young snake is enclosed in a transparent membrane, known as the pellicle, when born, from which it quickly escapes by thrusting with the head until the membrane breaks. Sometimes the pellicle is ruptured at birth. Both adder and smooth snake give birth to between about five and fourteen young, the actual number depending on the size of the mother. The adder has been known to give birth to more than fourteen young, but this is unusual. The average size litter for a fully grown adder or smooth snake is ten.

The young snakes are capable of living independently of their mother from the moment they are born, but they sometimes remain with her for a few days, basking beside her in the sunshine. Then one by one they wander off on exploratory rambles and do not return.

As with the young of most snakes the individuals in one litter vary in size, those of the smooth snake measuring at birth from between 15 and 18 cm in length, while newly born adders can measure anywhere between 15 and 20 cm, although 17 cm is an average length.

When first born smooth snakes are dark grey in colour, but after shedding their skins they are of a colour similar to that of the adult snake. Newly born adders are also greyish in colour, but after sloughing they emerge a dark brown.

Infant Snakes

Infant grass snakes are invariably more timid and alert to danger than either baby smooth snakes or baby adders. Young grass snakes that are handled immediately after hatching from the egg usually struggle frantically to escape, and when basking in captivity they will rush for cover the moment their cage is opened, or when some sudden vibration disturbs them. Baby adders are less easily alarmed, and I have frequently handled those born in captivity without their making any attempt to bite. Indeed I have seldom known infant adders to strike when picked up, which is surprising as the young adder is equipped with poison fangs when born, and can erect them in the manner of the adult, as can be seen when it yawns. Nor, as a rule, does the baby adder struggle violently to escape when picked up. I have in fact known them to coil up quite contentedly in the palm of my hand, obviously enjoying the warmth from it.

Newly born smooth snakes are equally as calm as, if not calmer in disposition, than young adders, and can sometimes be handled whilst feeding without their attempting to release their prey. I can recall on one occasion lifting a piece of bark on which a number of captive young smooth snakes were basking without their showing the slightest alarm. On replacing the bark on the ground most of the baby snakes continued to bask as though nothing untoward had occurred, although a couple departed unhurriedly into hiding.

CHAPTER FIVE

Feeding behaviour

Infrequency of meals – capturing prey – swallowing prey alive – fishing – constriction – battles with slow-worms – attacking with venom

UNLIKE most other animals snakes do not need to feed every day. One good meal will satisfy a snake for about a week, perhaps longer. At temperatures higher than normal snakes tend to feed more frequently, and during very warm weather I have known the grass snake to devour one fully grown frog every two to three days. But if necessary a snake can go without food for quite long periods, an ability that would enable it to remain alive when food is scarce long after other animals would have perished from hunger. An anaconda once survived in the Edinburgh Zoo for a whole year without food, and was eventually tempted to feed on a duck, after which it continued to feed regularly. There are numerous other cases on record of captive snakes going without food for long periods, and while many of these may have been due to the snakes being confined in cramped and unnatural surroundings, a refusal to feed sometimes results from a snake not having been offered the type of prey on which it is accustomed to feeding.

Just how long a snake can survive without food is largely dependent upon two things: its degree of activity, and the temperature of its surroundings. Some species of snake are very lethargic by nature and will spend hours lying motionless. Others are of a more restless disposition, and even when basking will constantly change their position. The heavy-bodied snakes, such as the pythons, boas, and the Puff Adder, are invariably the more lethargic, while the more slender species, tree snakes for example, tend to move about more. It is the lethargic snake that is able to stay alive the longest without food as it uses up less energy. However, as all snakes are cold-blooded their degree of activity is governed by the temperature of their environment, and at low temperatures even the most restless type of snake becomes less active, which enables it to conserve its bodily reserves so that it is able to survive longer without feeding. This is borne out by the hibernating snake, which emerges in the spring looking as plump and well fed as when hibernation began several months previously.

The way in which snakes obtain food varies with different species, and while there are some that lie in wait for their prey, most go in search of it. The majority of snakes have no means of either killing or subduing their prey and simply capture and swallow it alive. Snakes belonging to this category invariably feed on cold-blooded prey, such as fish and amphibians, which does not put up much of a struggle and which is also unable to retaliate when captured. Snakes that feed on prey that violently resists being eaten usually have some means of overpowering it. One method used is that of constriction, whereby prey is encircled by the snake's body and squeezed until it can no longer breathe. An even more efficient method of subduing prey employed by many different species of snakes is the injection of venom or poison. From the point of view of both the snake and its victim this latter method is by far the best. For the snake it means avoiding the risk of injury in a violent struggle with its prey, while for the latter death comes much more quickly than is the case with either constriction or swallowing alive.

Lacking the kind of teeth required to tear food apart the snake has no choice but to swallow its meals whole. There is nothing particularly unusual in this as there are other animals which do the same. But the snake is unique in that it is capable of swallowing entire animals bigger in girth than itself. This is made possible by the extreme elasticity of the snake's skin, plus the fact that the snake's jaws, unlike those of other animals, are not rigidly hinged at the back. This permits the lower jaw and the upper jaw to be widely separated so that animals considerably larger in size than the snake's own head can be swallowed.

Most harmless snakes possess only very small, recurved teeth, which are for the dual purpose of holding on to prey and assisting the passage of food through the mouth and into the gullet. Some harmless snakes, such as the smooth snake, also possess a few larger teeth at the back of the jaws which, by piercing the body of prey being swallowed alive, assist in its demise. The poisonous snakes have, in addition to the small recurved teeth just mentioned, longer, venom injecting fangs situated either at the front or the rear of the jaw.

When swallowing its prey the snake moves first one side of its jaw forward and then the other alternately, while at the same time both lower and upper jaw move forward independently of each other, so that the creature being swallowed is slowly drawn into the mouth, and from there on into the gullet. Once the meal has entered the snake's throat a muscular action assists in pushing it along the gullet and towards the stomach. Where the prey being swallowed is unusually long, as for example when a smooth snake swallows a slow-worm, an additional side to side undulating action of the body is brought into play, which helps to push the meal still more rapidly along the gullet.

To enable the snake to breathe whilst engaged in the lengthy process of swallowing a large meal, the windpipe or glottis is extended from its usual position in the throat out of the side of the mouth, and if one observes closely, the end of the glottis can be seen to open and close periodically as the swallowing of the meal proceeds. Small prey which can be easily and quickly swallowed does not necessitate protrusion of the glottis.

Digestion begins as soon as the meal reaches the stomach, although it sometimes happens that the whole of a very large meal cannot be accommodated by the stomach, and part of it has to remain in the gullet while the other part is being digested. The snake's digestive secretions are very powerful and even the skeleton of the creature that has been swallowed is eventually dissolved, although fur, feathers, and teeth are more resistant to the digestive processes and are often passed out of the body in the faeces.

Solid waste matter passes from the intestine into the rectum, and from there into the cloaca, into which the ureter, or urinary duct, also opens. From the cloaca both solid and fluid waste matter are excreted from a single opening at the base of the tail in the form of semi-liquid faeces which are either white or yellow in colour. When excreting the snake usually lifts either the base or the whole of the tail from the ground, and after having excreted may sometimes move about for a while with the tail still raised.

When alarmed the grass snake will often void its excreta mixed with a secretion from the anal gland which possesses a powerful and lasting odour. The smooth snake occasionally behaves in the same way, whereas the adder rarely does so.

Most non-venomous snakes invariably seize their prey by the side of the body close to the head, and for very good reasons. One is that prey grasped in this way cannot easily turn and bite, but that is not all. The snake could make a frontal attack and grasp its prey by the head, but the target presented by a head pointing towards it would be a narrow one and the snake's sudden lunge to grasp it in its jaws might easily miss and the intended victim escape. An attack from the side is likely to be more successful since if the snake's prey were to make a sudden dash for freedom the snake still has the chance of taking hold of it across the body. Furthermore if the snake were to face its prey and try to grasp it by the snout it might retaliate and grasp the snake's head in its own jaws. By taking hold of its prey by the neck the snake can avoid being bitten, and by working its jaws sideways without releasing its hold it can keep its victim's jaws closed while it engulfs the head nose first.

As a rule snakes prefer to swallow their prey head first, and there are two good reasons why they choose to do so. Firstly, prey with legs is more easily swallowed head first because the legs fold naturally backwards against the sides of the body. If swallowed rear end first the limbs tend to stick out at an awkward angle to the body thus making the creature more difficult to engulf. Secondly, if the creature being swallowed possesses teeth, and is still alive when swallowed, it would have more opportunity to bite the snake if swallowed tail first.

When swallowed head first the copious flow of saliva the snake produces soon blocks the nasal passages and mouth of the victim and quickly suffocates it. At least that is what usually happens, although there are cases on record where the snake itself has been killed by its victim biting it from within. Frogs and toads are not easily suffocated in this way as they are able to survive on a very limited supply of oxygen, and they have been known to remain alive for several hours inside a snake. I can confirm this, having once captured a grass snake in Ashtead Forest in Surrey that disgorged a live toad. Apart from a few small punctures in its body caused by the snake's teeth it was completely unharmed. I placed both grass snake and toad together in my collecting bag, thinking that the snake would not attempt to re-swallow the toad under such conditions, but I was mistaken. For on looking into the bag or reaching home I discovered that the toad had disappeared, and that the bulge that had been in the snake's body before it disgorged the toad had reappeared.

Most snakes would not attempt to feed under the conditions just described, nor would they do so while being watched, but they are often reluctant to relinquish a meal once they have begun to swallow it. They may release it immediately on being disturbed by an intruder, but they will invariably return later to swallow it, and I have known them to do so in full view of an observer. The grass snake is particularly persistent in this respect.

Perhaps the most surprising thing about the British species of snake is that although only three in number they each use one of the three different methods employed by snakes to overpower their prey. Firstly we have the grass snake which captures and swallows its prey alive. Secondly there is the smooth snake which is a constrictor, and thirdly the adder which kills its prey with venom.

Grass Snake

The grass snake's list of prey includes frogs, toads, newts, fish, tadpoles, mice, and birds. Although the frog is an agile creature capable of moving quite rapidly in a series of prodigious bounds it is not difficult for the grass snake to capture. As a rule the grass snake takes hold of the frog by the side of the body and then works its jaws sideways to swallow it head first. Often, however, the grass snake captures the fleeing frog by one hind leg, and rather than risk losing a meal by releasing its hold on the leg to grasp the frog's body, the snake swallows the frog from the rear end. It does this by first swallowing one hind leg, and then the remainder of the frog. When the snake's jaws reach the frog's body the other hind leg becomes extended forward, so that it lies alongside the frog's body with the foot reaching beyond the nose. Having once been taken hold of by a snake the frog offers little or no resistance to being swallowed. It appears resigned to its fate and makes not the slightest protest.

It has frequently been stated that the grass snake does not normally feed on toads and will only take them if very hungry. My own observations do not confirm this, as I have on several occasions captured grass snakes that have disgorged the remains of a common toad. I have also known captive grass snakes to feed readily on toads, and often when there were frogs available. The common toad's reaction to the approach of a grass snake is quite different from that of the frog. Being less agile than the frog it has little chance of getting away from a snake once the latter has found it. It seems to realise this and instead of trying to take refuge in flight it resorts to strategy to avoid being eaten. If the snake does not approach too close the toad will remain perfectly still, only the pulsating of its throat, and the occasional blink of an eye, revealing that it is alive at all. By simply remaining motionless the toad can sometimes escape being eaten, for snakes rely on movement of their prey to inform them of its whereabouts. The eyesight of most snakes is very poor, and most of them seem unable to recognise their prey even from a few inches away unless it moves. This is the reason why a snake will often lie and stare for minutes on end at something that has attracted its attention and then suddenly become immobile. If no further movement occurs the snake will often lose interest and move away. It is this habit of gazing fixedly at prey that has led people to believe that snakes can prevent their prey from escaping by hypnotising it, but such is not the case, for snake's possess no such power. It is simply that many wild animals know that by keeping absolutely motionless they can often escape being eaten by a predator, whether it be a snake or some other carnivore. Snakes, however, seem to possess more patience than most animals, and before moving away a snake will spend a long time staring at a creature that remains unmoving. Sometimes the snake's patience outlasts that of its prey, which, on finding that the snake does not move, will often try to steal quietly away, or make a sudden dash for cover. Whichever mode of departure it chooses the snake is quick to spot the movement and instantly attacks.

The grass snake does not always wait for its prey to make the first move and, after watching it closely for some time, may approach to make a closer inspection. If the creature still does not move the snake will play its tongue over it to discover if it is edible. Having established that it is the snake may, without waiting for a sign of life on the part of its prey, simply take hold of it and commence to swallow it. But the common toad does not usually wait for the snake's tongue to touch its body, and on seeing the snake approaching it inflates its body with air, while at the same time extending its legs until it is standing on tiptoe with its body clear of the ground. By doing this it makes itself appear much larger than it really is, which may deter the snake from attempting to swallow it. For a snake knows its own capacity and, as a rule, will not attempt to swallow too large a meal. However, the snake is not instantly put off by the toad's sudden increase in size, and proceeds to inspect it at close range from all angles. If the snake approaches the toad from one side the toad tilts its body sideways towards it. If the toad is approached from behind by the snake it raises the forepart of its body by straightening its front legs, at the same time lowering its posterior; whilst if the snake makes a frontal approach the toad lowers the forepart of its body until its nose is almost touching the ground, and extends its hind legs to raise its rear end. By behaving in this way the toad presents the largest area of its body to the snake, thus making itself appear bigger than it actually is. The snake is frequently fooled by the toad's behaviour, and goes off in search of smaller prey. It is not always so easily discouraged, however, and will sometimes move a short distance away where it lies watching the toad. As soon as the snake moves away the toad lowers its belly to the ground and sits quite still for a while, its body slowly deflating. If the snake does not approach again the toad will usually make off. Immediately it starts to move the snake's interest is once more aroused, and after a moment's pause it rushes after the departing toad. The rapid approach of the snake sometimes takes the toad by surprise, particularly where the snake advances on it from behind, and before the toad has time to adopt its defensive pose it is seized by the snake. But even at this stage the toad may still escape being swallowed again by inflating itself with air, thus making itself too large for the grass snake's jaws to engulf.

Normally when a toad adopts the defensive attitude just described it does so fairly slowly, rising gradually and shakily until its body is raised as high as the fully extended legs can take it. I can, however, recall an occasion when a toad I was watching did this so suddenly that the snake that was gliding over its back was forcibly thrown a distance of several inches. The snake landed on its back but quickly righted itself and made towards the toad again. The snake did not appear to be interested in the toad as prey, but was merely exploring its enclosure in an attempt to find a way of escape. When it again passed over the toad's back the same thing was repeated and the snake was thrown once more.

Newts are just as readily eaten as frogs and toads by the grass snake, and it has no difficulty in capturing them either on land or in the water. A fully-grown frog or toad will satisfy an adult grass snake for about a week, whereas several newts are required to satisfy its appetite for the same length of time. Fish are equally as acceptable as food, but as they are quicker in movement than amphibians many grass snakes seem to have difficulty in capturing them, although a few are quite adept at it. I have known some grass snakes that could capture a fish with one rapid lunge, but the majority of grass snakes often make several clumsy attempts before being successful. Those that can do it easily have most probably been accustomed to feeding on fish, whereas those grass snakes that have perhaps fed almost exclusively on amphibians have not developed reactions fast enough to cope with such prey. My own observations on the grass snake's efforts at capturing fish have led me to believe that fish do not figure prominently in the diet of the majority of grass snakes. For although a snake can move fairly swiftly in water, fish dart about at a much greater speed, which makes them difficult for the snake to capture. Nevertheless, most grass snakes will feed on both fish and amphibians, and it has not been my experience that a snake accustomed to feeding on frogs will not feed on fish, or the other way about.

When trying to capture a fish the grass snake appears to be handicapped to a far greater extent by its poor eyesight than when it is hunting amphibians on land. A grass snake peering into a pool from above the surface seems quite unable to see fish that may be swimming about a few inches below. Even when the fish approach the surface of the water the snake still appears unable to see them if they are moving slowly, and its attention is only attracted to them by the darting actions that fish often indulge in. As a rule the grass snake has the most success at fish catching when it lies with its head overhanging a pool in which fish are swimming. It appears to be able to see them a little more clearly from that position than when it actually enters the water to hunt for them. A fish moving slowly can often pass within an inch of a grass snake's eye when the snake is swimming underwater and not be recognised as prey. But should the fish dart past suddenly, or jump about near the surface, then the snake's interest is at once aroused and it rushes in pursuit. Sometimes when a grass snake has made an unsuccessful lunge at a passing fish it will swim around with its jaws open in readiness to snap in the direction of any sudden disturbance in the water that would indicate the presence of a fish.

It seems probable that fish are only caught on occasion, such as when a grass snake hunting for frogs around the edge of a pool happens to catch sight of a fish, or shoal of fish, darting about near the surface. Having sighted a fish the snake watches it intently, and when the fish darts close by the snake plunges its head and the forepart of its body beneath the surface of the water. Its sudden lunge, although very rapid, is not always successful, but as fish small enough for the grass snake to swallow usually swim in shoals, the snake invariably has more than one chance to make a capture. If its sudden attack frightens the fish away the snake will continue watching until the fish reappear, or until it gets tired of watching. When successful in making a capture the snake does not usually snatch the fish quickly from the water, but in direct contrast to its sudden attack, it lifts the often struggling creature quite slowly from its element and carries it on to the bank to be devoured. If the snake makes a capture while swimming it will often pause to secure a firm grip on the fish before bringing it ashore to be swallowed.

A fully-grown grass snake can swallow two or three fish of about 8 cm in length one after the other, and I have known grass snakes to devour gudgeon measuring as much as 11 cm in length. Because of their streamlined shape fish are more easily swallowed than either frogs or toads, and I have seen a grass snake 76 cm in length swallow a gudgeon 10 cm in length in under twenty seconds.

Tadpoles are readily eaten by both young and adult grass snakes, and I have known two adult grass snakes to devour between them two dozen tadpoles of the common toad within a very short time.

Captive grass snakes have been known to feed on dead mice, and it is possible that a grass snake living wild might eat either a mouse or vole that had recently died if it chanced to find one. But whether the grass snake would tackle a live mouse or vole is doubtful, particularly as it has no means of defence against animals capable of biting. For the same reason it is unlikely that the grass snake would attempt the capture of a fully-fledged bird. On being seized the bird would almost certainly retaliate by pecking, with possibly disastrous consequences for the snake. However, in the summer of 1943 I captured in Ashtead Forest in Surrey a large grass snake that disgorged the remains of a fully-fledged bird about the size of a sparrow. It is possible that the bird approached close to the snake in an attempt to drive it away from its nest, and was caught and swallowed by the snake. It is more probable that the bird was a young, fully-fledged bird that was not yet able to fly properly. An even more likely explanation is that the snake found the bird already dead. Snakes will feed on dead prey if it is reasonably fresh, as I have myself discovered, but I will deal more fully with this aspect of the feeding habits of snakes later in this chapter.

Although it is unusual for the grass snake to capture and devour warm-blooded animals, whether furred or feathered, it does occasionally happen, as I discovered in July 1952 when on Arbrook Common in Surrey I caught a grass snake that brought up three unfledged baby birds. The snake was captured at about 10 a.m., and had obviously swallowed the baby birds only shortly beforehand, as digestion had not even begun. The infant birds were very small, and were possibly the young of a robin or some other bird of about that size. The nest from which they were taken may have been on or near the ground, which would make their capture a simple matter for the grass snake. It is possible, however, that the snake had climbed into the branches of a bush to catch the rays of the rising sun, and in doing so chanced upon the nest containing the young birds. Whichever way the grass snake came upon them it seems unlikely that they were dead when found. This, I consider, is sufficient proof that the grass snake does not always confine itself to a diet of fish and amphibians, but will occasionally devour small, warm-blooded creatures that are incapable of retaliating when seized. But strangely enough the grass snake will never, to my knowledge at least, attempt to feed on lizards, which are cold-blooded like its normal prey. On several occasions I have seen a common lizard pass close to a basking grass snake which showed not the slightest interest. Occasionally a grass snake will make as though to pursue a passing lizard, and then stop immediately it recognises the creature.

Smooth Snake

The smooth snake's method of capturing prey is a little more advanced than that of the grass snake, since it can prevent its victim from escaping by encircling it with its body. It can also, if necessary, kill its prey by constriction in the manner of the giant pythons and boas. The smooth snake feeds almost exclusively on the common lizard and the slow-worm, although it occasionally devours mice and voles. When hunting for prey it has either sighted or scented, the smooth snake seems to take on the behaviour of a lizard. Its movements become quick and agitated as it searches feverishly about, and on catching sight of its intended victim it will, if sufficiently close, make a sudden dash forward to capture it. Should it sight its prey from some little distance away it will make a stealthy approach, edging very slowly forward, and all the while drawing up the forepart of the body into an S-shaped loop in readiness for the sudden lunge it will make to capture its prey. It is not always successful in capturing an agile creature like the common lizard at its first attempt, the lizard often succeeding by its quickness in escaping the smooth snake's fast attack.

As with most animals in the presence of a snake, the lizard endeavours to keep as still as possible, for it knows instinctively that by doing so it is less likely to attract the snake's attention. It remains immobile for as long as it can, but as the snake draws nearer either its nerve breaks, or it realises that the snake is already aware of its presence, and so it makes a last second dash for safety. Sometimes it waits just a little too long and is seized by the snake.

Usually the smooth snake endeavours to take hold of the lizard by the side of the body close to the head, from which position it moves its jaws sideways until the head is reached, whereupon the swallowing of the lizard commences. But because of the lizard's agility the snake is not always successful in its aim, and its jaws close around the lizard's tail. This appendage is designed to fracture easily to enable the lizard to escape its enemies, and I have seen a common lizard's tail break off when taken hold of by a smooth snake. The smooth snake promptly swallowed the still wriggling tail, and then went off in the direction taken by the lizard. It was obviously not fooled by the lizard's little trick, and seemed to know that there should be more to a lizard than just a tail, and it was not long in finding and devouring the owner.

When successful in capturing a lizard the smooth snake may swallow it without attempting to restrain its struggles by throwing its coils around it. If, however, the lizard is a large one and struggles violently the smooth snake will often throw a coil or two around it to immobilise it while swallowing proceeds. Sometimes the lizard will attempt to bite its captor, and in such cases the smooth snake may apply full constriction until the lizard is gasping for breath and no longer has any inclination to retaliate. Small lizards that cannot be held within the snake's coils are often pressed against the ground to prevent their escape and also to assist their passage into the snake's throat.

The smooth snake preys equally as readily on the slow-worm, and will tackle quite large specimens, sometimes more than half its own length in size. Swallowing a meal of these dimensions is a lengthy business, and can take up to two hours. I once timed a smooth snake measuring 61 cm in length that was swallowing a slow-worm 30 cm long, and found that it took one hour and twenty-five minutes. Another smooth snake measuring 55 cm took one hour and fifty-one minutes to dispose of a fully-grown slow-worm.

As it is less agile than the common lizard the slow-worm is

more easily captured by the smooth snake, although, being larger and stronger, it is not so easily subdued. Some slow-worms struggle more violently than others when captured by a smooth snake, but no matter how vigorously they resist they seldom escape. Although snake-like in form the slow-worm seems to be instantly recognised by the smooth snake as prey, and is attacked without hesitation. Invariably the legless lizard is grasped in the snake's jaws just behind the head. If little resistance is offered the smooth snake may simply throw a few loose coils around its victim to hold it still whilst it swallows it alive. Sometimes, however, a violent struggle ensues with smooth snake and slow-worm writhing and rolling in a life and death battle. The two of them appear inextricably entangled, and at first it appears doubtful whether the snake can control its struggling victim sufficiently to be able to bring its powers of constriction into operation. But with surprising suddenness the smooth snake gains the advantage, and emerges from the tangle with its powerful body encircling that of the slow-worm in a series of close, tight coils, rather like those of a spring. Its hold on the slow-worm is an extremely tight one as is illustrated by the way the slow-worm's body bulges where it emerges from the snake's coils. This is true constriction, and if applied to a warm-blooded animal it would kill very quickly by restricting the lung action.

Contrary to popular belief constricting snakes do not crush their victims to death, but simply draw their coils tighter each time the victim exhales, thereby gradually restricting the action of its lungs until death eventually results from suffocation. But the death of a cold-blooded animal is not easily brought about in this way, as it can survive for a much longer period on a very limited supply of oxygen than can a warm-blooded creature. Nevertheless, I have seen a slow-worm constricted by a smooth snake for a period of fifteen minutes become limp and with mouth agape as though quite dead.

Having either killed its victim or reduced its struggles the smooth snake then moves its jaws sideways until the head is reached, and the slow process of swallowing it then begins.

When trying to swallow a slow-worm that has not been constricted and still continues to struggle, the smooth snake sometimes gets into some very awkward and uncomfortable looking positions. I have seen a smooth snake with its head upside down and its neck twisted in a manner that appeared quite painful, as it endeavoured to engulf the head of a still struggling slow-worm. To avoid the risk of its victim escaping while the business of swallowing it proceeds, the smooth snake remains coiled around it, drawing its victim through its coils until a large part of it has disappeared down its gullet, when the hold is released.

A habit sometimes indulged in by the smooth snake when feeding is that of anchoring its tail around some inanimate object, such as a stone or the stem of a plant. As the smooth snake's tail is strongly prehensile, a hold of this kind may assist the snake to apply more pressure when constricting prey, although it frequently behaves in this way when swallowing prey which has not first been constricted. On one occasion, whilst watching a smooth snake swallowing a common lizard, I saw the snake grope around with its tail for some kind of anchorage. The tail came into contact with another smooth snake that was basking nearby, and immediately the groping tail fastened on to it. So engrossed was the feeding snake in holding on to the struggling lizard that it was seemingly unaware of what its tail had caught hold of, and whilst it endeavoured to tighten its grip on the lizard, it pulled hard with its tail on what it apparently believed was a firm anchorage, with the result that part of the other snake's body was lifted off the ground. The basking snake did not take kindly to having its siesta so rudely disturbed and promptly fastened its jaws on to the posterior half of the feeding snake and wrapped its body around it. For about thirty seconds it chewed its jaws forward until it had a firm grip on the feeding snake's body, and then suddenly seemed to realise that it was biting a fellow snake and immediately released its hold and moved away.

All the time this was going on the snake that was busily occupied in devouring the lizard just carried on as though nothing was amiss. Certainly it did not appear to feel the bite of the other snake, for it made no effort to move that part of its body that was being bitten, nor did it show the slightest sign of resentment. Indeed, a feeding smooth snake seems to be oblivious to everything but the task in hand, and can often be picked up without making it release its prey.

Although a smooth snake is quick to spot a moving lizard at a distance of up to two feet away, its eyesight is poor, and it seems unable to recognise prey that remains stationary until it gets within a few inches of it, and sometimes not even then. This was clearly illustrated to me on one occasion when I was watching a smooth snake hunting for a lizard that had been successful in avoiding capture when the snake first attacked. For a while the snake searched about in an agitated manner trying to find the elusive lizard, but on meeting with no success it settled down to watch and wait for the lizard to make a move that would reveal its whereabouts. It chose to lie close to a young smooth snake much smaller than itself that was quietly basking, and no sooner had it settled itself when the smaller snake decided to move. It travelled only a short distance and then stopped, but at this sign of movement the bigger snake was instantly on the alert with head raised, and when the smaller snake moved again the big snake dashed forward and fastened its jaw on to its tail. Startled by this unexpected attack the young snake wriggled frantically to free itself, but the bigger snake hung on, chewing its jaws forward on to what it must have believed to be the lizard it had previously been hunting. After about fifteen seconds it seemed to realise its mistake and released its hold, whereupon the little snake glided quickly beneath the shelter of a nearby rock.

The incident just related gives some idea of how excited or keyed up a snake becomes when hunting its prey. Even though it may settle down quietly to wait for the prey it has been hunting to reveal itself it is still very much alert, and any slight movement near to it will rouse it instantly to a state of tense anticipation.

Although I have read that the adult smooth snake is

partial to eating young snakes, either its own or those of other species, I have never known this to happen. However, the cannibalistic tendencies of the smooth snake was on one occasion illustrated to me by a young smooth snake, only six weeks old, that I found trying to swallow another young smooth snake. They both belonged to the same litter and when first discovered the head and forepart of one baby smooth snake was inside the gullet of the other. On being disturbed the young smooth snake made no attempt to relinquish its meal, but on being picked up it disgorged it, and the baby smooth snake that had been partly swallowed was found to be dead, but completed undigested. It was, therefore, evident that the swallowing of it had only recently taken place. One might think that this was obvious as only part of the baby snake had been swallowed. But it sometimes happens that a snake will attempt to swallow a meal too large for it to completely engulf, and rather than disgorge what it has already swallowed, it allows part of the meal to remain outside its jaws until the part inside its body has been digested. In this way the meal can be digested a little at a time until all of it has finally disappeared down the snake's gullet.

Whether the baby smooth snake had been swallowed after having died from some natural cause, or swallowed alive, was impossible to determine. But this incident does illustrate that the smooth snake is capable of devouring its own kind, although from my past experience of this snake's habits I am inclined to believe that it is of very rare occurrence. I must hasten to add that the incident related took place in captivity, and the same litter of young smooth snakes at one time shared an enclosure with a number of baby grass snakes, none of which was ever attacked or eaten by the smooth snakes. Nor did the young smooth snakes take much interest in a recently dead infant grass snake that was placed among them. Two of the baby smooth snakes inspected the dead snake, but neither of them made any attempt to devour it.

The usual prey of the infant smooth snake is the young of common lizards and slow-worms, which it attacks and overpowers in the same manner as the adult snake. I have seen a five weeks old smooth snake wrap itself around a baby slow-worm and swallow it after a vigorous struggle. The infant snake did not appear to be constricting the slow-worm, but used its coils, somewhat unsuccessfully to hold it still.

Although alleged to feed on insects I have never found baby smooth snakes to take prey of that kind, although I have given them the opportunity to feed on worms, slugs, and various kinds of insects.

ADDER

Of the three British species of snake the adder enjoys the most varied diet. It feeds readily on common lizards, slow-worms, voles, mice, and will also devour both unfledged and fully-fledged birds when it gets the chance, which probably is not very often. It has also been known to eat amphibians, although I have never, myself, recorded such prey being taken. Many adders show a preference for common lizards, but most will devour mice and voles when lizards are not available. Indeed, I am inclined to believe that the adder will devour almost any small animal, either cold-blooded or warm-blooded, when it is hungry. Being able to quickly overpower its prey with an injection of venom it runs little risk of being injured by either rodents or small birds, and so it can venture to attack, with the purpose of devouring, almost any small creature that crosses its path.

Like the grass snake and the smooth snake the adder usually goes in search of its prey, but should either lizard or mouse venture near it while it lies basking it will, if sufficiently hungry, give chase. It can move with surprising speed when in pursuit of prey, and can strike successfully at it whilst on the move, as I have seen it do on more than one occasion. If it loses sight of its intended victim it will scent it out, using its tongue for this purpose. When rush tactics fail, it will stalk its prey with all the stealth of a cat, drawing gradually closer to it until within striking distance. It then waits, with the forepart of its body drawn up into an S-shaped loop, for the animal it is hunting to make a move, whereupon it strikes with lightning speed.

There is considerable force behind an adder's strike and I have seen a common lizard knocked off balance when struck at by an adder. Quickly recovering from the blow the lizard continues its frantic dash for cover and is soon lost to sight. The adder makes no attempt to go in pursuit, but waits calmly for perhaps two or three minutes before starting to trail its victim to where it lies either dead or dying. It does this largely by scent, and to watch an adder following the path taken by its stricken prey is rather like watching a bloodhound at work. The adder first lifts its head higher than usual above the ground and while remaining absolutely motionless in this position it waves its forked tongue slowly in an attempt to gather scent particles from the air. It then lowers its head and unerringly takes the right direction, stopping frequently to touch with its tongue the herbage against which the lizard has brushed in its panic-stricken flight, and every now and then pausing to press its nose to the tiny drops of blood left behind by the lizard. Eventually it reaches the last hiding-place of its victim, and taking hold of the lizard by the snout it begins the slow process of swallowing it.

For the adder there is none of the bother of having to grab its prey quickly, and then work its jaws slowly towards the head before it can be swallowed. Its prey, having been killed within one to three minutes by the venom injected into it, cannot struggle to escape, and so the adder is able to take hold of it by the head and swallow it without delay, and also without the risk of losing it. Furthermore, as the action of the venom is to break down tissue and blood the adder's meal is already partly digested before it is swallowed.

Proof that the adder will feed on small birds when it gets the opportunity was provided by a female adder that I captured on Arbrook Common in Surrey. Shortly after catching the adder at about 10 a.m. it disgorged the almost wholly undigested body of a fully-fledged bird about the size of a sparrow. The adder measured 22 cm in length, and for a snake of this size a sparrow would be a large meal, which would take some time to swallow. How the adder came by such a meal is anyone's guess. The adder is not much of a climber, and so it hardly is likely to have captured the bird on its nest above ground level. It is possible, of course, that the bird was a ground nesting species, in which case it would not be difficult for the adder to attack if it sat tight on its nest when the reptile approached. A more likely explanation is that the bird attempted to drive the adder away from the vicinity of its nest and the adder struck at it in self-defence. Having killed the bird the adder would automatically try to swallow it. For with the adder, feeding is almost like a reflex action after it has used its poison fangs, although only, of course, in the case of creatures small enough for it to devour. An example of this was once illustrated to me by a captive adder that became irritated by a common lizard that repeatedly approached close to it while it was basking, and then dashed off in fright. In its hurried flight the lizard would sometimes run over the adder's body, and on one occasion it trod on the adder's head. This seemed to alarm the adder, and when the lizard came rushing past again the adder struck at it with its poison fangs, more in self-defence it seemed than for any other reason. The adder had certainly not been hunting for food, and yet after having struck at the lizard it followed up its action by swallowing the lizard. It was as if the adder's feeding mechanism had been triggered off by having used its poison fangs, or perhaps by the taste of blood. Prior to this incident the adder had shown not the slightest desire to feed whilst in captivity.

The young adder feeds on the young of the creatures it will prey upon when adult, and I have no experience of it ever feeding on insects or small invertebrates. I have, however, recorded among my notes on adder behaviour, two cases of young adders no more than 25 cm in length swallowing fully-grown common lizards.

Although the adder's eye has an elliptical iris, which is normally associated with nocturnal behaviour, the adder is in fact chiefly diurnal in its habits, and its main functions, such as feeding, mating, and sloughing are all performed during daylight hours. However, Dr Malcolm Smith, states in his book *British Amphibians and Reptiles* that he has known some captive adders to feed only at night. I have never observed this nocturnal behaviour of the adder, but as adders are often reluctant to feed in captivity observations on their feeding habits are not easily made. On the few occasions when I have witnessed captive adders feeding it has occurred when there was sufficient light for them to be able to see their prey. Nevertheless, I have known both grass snakes and smooth snakes to feed at night, and so I must assume that the adder may possibly do so at times. But as a rule all three British species of snake feed mainly during daylight hours. It is possible that on particularly warm summer nights snakes may be tempted to go hunting for food, but that under more normal conditions, that is when the night time temperature falls considerably lower than that of the day, snakes would feed only during the day. I have found with captive snakes that they invariably retire under cover at night, and do not emerge from their hiding-places until sunrise the following day.

Although I have stated earlier in this chapter that snakes have poor eyesight, and that, as a result, they rely on movement of their prey not only to be able to locate it, but also to stimulate their desire to feed, there are occasions when snakes will take dead food, and this is by no means an infrequent occurrence. Most grass snakes will, I have found, eat dead fish that is still fresh. If the fish has been dead for several hours and is beginning to smell, although perhaps only slightly, the grass snake will have nothing to do with it. However, one grass snake I had in captivity ate a small gudgeon that had been dead for thirteen hours, but prey that has been dead that long is not usually taken.

That some snakes will devour dead prey readily, and not merely resort to doing so when desperately hungry, is clearly proved by an observation I once made while trying to photograph a grass snake swallowing a fish. A dead fish of about

16. Having slit the tough skin of the egg with its egg-tooth the infant grass snake may spend several hours with only its head showing before emerging completely.

The egg-tooth can be quite clearly seen in this photograph as a small projection on the tip of the nose.

17. Smooth snake basking with young. Although able to survive independently of their mother from the moment they are born the young snakes often choose to remain with her for a few days.

18. When searching for food the grass snake pokes and peers everywhere. Its tongue flickers constantly in and out, and the moment it picks up the scent of a frog its movements become quicker and more agitated.

19. Common toad adopting a defensive attitude to avoid being eaten. The grass snake seems unable to decide whether or not this toad is too large to be swallowed and makes a closer inspection, playing its tongue over the toad's back. The toad tilts towards the snake.

20. If approached from behind the toad lowers its hindquarters and raises the forepart of its body.

21. The toad raises itself as high as it can on all four legs and tilts its back towards the grass snake. By making itself appear much larger than it really is it may escape being eaten by the snake.

22. This photograph dispels the widely held belief that the grass snake will not devour the common toad because of the poison which exudes from its skin. The grass snake in this picture swallowed the toad and suffered no harmful after effects.

23. Grass snake swallowing common toad. Prey is normally swallowed head first, but in this instance the toad has been swallowed hind feet first. This is a slower and more laborious task for the snake because the legs tend to stick out at awkward angles, thus making swallowing more difficult.

24. The toad's forelegs have become folded across its chest during the process of being swallowed by the snake. In this photograph one of the toad's elbows is seen creating an uncomfortable bulge just behind the neck of the snake.

25. A grass snake makes an unsuccessful lunge beneath the surface of the water to capture a fish. With a quick flip of its tail the fish neatly avoids being caught.

26. As the grass snake closes in on the fish its jaws open wider.

27. While still submerged the grass snake works its jaws sideways until it has the head of the fish engulfed before bringing it to the surface. This does not always happen, and more frequently the snake lifts the fish from the water while still holding it across the body.

28. Having captured the fish the grass snake usually carries it on to land to be swallowed.

29. Without releasing its hold the snake works its jaws sideways so that it can swallow the fish head first.

30. Another five seconds have passed and the fish is on the point of disappearing inside the snake's jaws. Total time taken to swallow fish – twenty seconds.

31. Although found most frequently in dry localities the adder is not averse to entering water. The adder in this photograph was seen to enter a pool of water on several occasions and explore beneath the surface.

9 cm in length was placed on a log about a foot from where a captive grass snake measuring 71 cm in length lay basking. Half an hour passed and the snake continued to bask, showing no interest whatsoever in the dead fish. The dead fish was then removed and placed in a small water container, and after a few minutes the snake came to take a look. It raised its head over the edge of the water bowl and played its tongue for some time over the dead fish, which was floating upside down near the surface of the water. The snake then moved away and began to bask on a log close by. It seemed that either the snake found the fish not to its liking, or that it was not hungry, having only shortly beforehand devoured two smaller fish. As for myself I was beginning to despair of getting the picture I wanted, but I decided to make one more attempt to arouse the snake's interest in the fish. Lifting it from the water bowl I placed it on the log about 15 cm from the snake's nose. After lying unmoving for about ten minutes gazing in the direction of the fish the grass snake at last approached it and, after a brief inspection, seized it near the head and began to swallow it. Within twenty seconds the fish had vanished from sight inside the snake's gullet. I have used the word 'seized' in its fullest sense, since the snake did not take hold of the fish slowly, but grasped it quickly in its jaws as though it expected the fish to try and escape.

In addition to eating prey that has been found dead, a snake will often re-swallow a meal that it has disgorged dead when disturbed by an intruder shortly after having fed. Only prey that has been disgorged within a short time of being swallowed is likely to be re-swallowed by the snake. Prey that has been disgorged partly digested is never, in my experience at least, swallowed a second time by the snake.

Some animals seem to be able to survive being completely swallowed alive by a snake, as, for example, the toad mentioned earlier in this chapter, which was disgorged very much alive by a grass snake, although it must have just about reached the snake's stomach, judging by the position of the bulge in the snake's body before the toad was disgorged. Other observers have given accounts of frogs being disgorged alive, and I have seen a grass snake disgorge a frog that had disappeared from sight down its throat. A large bulge just behind the snake's neck revealed the position of the frog inside the snake's body, and on being picked up the snake brought up the frog, which was still alive and apparently unharmed. After resting quietly for a few minutes the frog seemed none the worse for having been swallowed.

I have known several cases where a frog that has been partly swallowed, its head having reached the snake's throat, has been released by a grass snake that was disturbed whilst feeding. On being released the frog has often bounded away as though unaffected by the experience.

Many naturalists have stated that the frog when captured by a snake seems paralysed with fear, as it offers no resistance whatsoever to being swallowed. My own observations do not altogether support this view. A frog when taken hold of by the body may kick with its legs, but does not usually struggle violently. However, if the snake seizes the frog by one leg, as often happens, the frog invariably makes vigorous attempts to escape, and sometimes succeeds.

Like the grass snake the smooth snake will also, on occasion, take prey that is already dead, as I discovered in September 1965 when I recorded a most unusual incident in the behaviour of some smooth snakes I had captive at that time. In my notebook I made an entry to the effect that two fully-grown wall lizards had been placed in an enclosure containing four smooth snakes, and that I had observed one of the largest smooth snakes, measuring 61 cm in length, devouring both lizards one after the other, and taking twenty minutes to swallow each one. During the night both lizards were disgorged completely undigested. Being curious to discover whether or not the lizards would be re-swallowed I allowed them to remain in the enclosure for the whole of the following day. When evening came the dead lizards still remained in the spot where they had been disgorged, but that night the snake that had disgorged them re-swallowed one of them, as was evident by the bulge that was to be seen in its body the following morning. In other words the snake had re-swallowed a meal that it had disgorged twenty-four hours previously. This in itself was remarkable, as snakes do not normally take food that has been dead for so long, but even more surprising events were yet to take place. Another twenty-four hours later the snake again disgorged the lizard during the hours of darkness. But whilst the bulge had vanished from the snake's body there was no sign of the lizard. I then noticed that one of the smaller smooth snakes now had a very large bulge in its middle. There was no doubt that the smaller snake had swallowed the very same lizard disgorged by the bigger snake, as the other dead lizard still lay in the very same spot where it had been disgorged forty-eight hours earlier.

The extraordinary behaviour of these two smooth snakes was something that I would have found difficult to believe had I not witnessed it for myself. It was surprising enough that a snake would devour an animal that had been dead for twenty-four hours, but for a snake to eat an animal that had been dead for twice that long, and after it had been disgorged by another snake, was quite astounding. It might be of interest to the reader to know that the smaller snake also disgorged its second-hand meal, so that in the end none of the snakes had the wall lizards for a meal.

Although I cannot claim to have witnessed an adder taking food that it had not killed itself, I see no reason why it should not do so, particularly as it is accustomed to eating animals that offer no resistance to being swallowed, having first been killed by an injection of venom.

Accepting that all three British species of snake are chiefly diurnal it can be safely said that a hungry snake will feed at any time throughout the day, providing the weather is sufficiently warm. However, from my own observations I have found that feeding takes place more often in the early morning during the warmer months of the year. As stated earlier in this chapter, a snake must be sufficiently warm before it can feel inclined to eat. At a low temperature, and by this I mean any temperature below 15° C, British snakes are disinclined to feed, which is the reason why they invariably spend an hour or so basking just after sunrise before going in search of food. During normal summer weather all three British snakes will emerge to bask as soon as the sun's rays reach their hiding-place in the early morning. But in very warm weather, when night time temperatures may not drop below 15° C, snakes do not always wait for the sun to warm the air around their hide-away before they emerge. Being already warm enough to be active they emerge as soon as it is daylight and make for a spot that has already been warmed by the sun's rays. If surrounding trees and bushes screen the low rays of the rising sun the more agile grass snake will sometimes climb into a low bush to catch the first rays of the sun.

When night temperatures are lower snakes do not come out of hiding soon after daybreak, but wait until they feel the temperature of their surroundings begin to rise. Then, in a somewhat lethargic state, they slowly emerge, and after luxuriating in the sun for a while, go off in search of a meal. If a snake does not feel inclined to feed it may bask in the sun until the heat becomes too much for it, whereupon it retires to a cooler spot. During very hot weather the temperature may remain quite high even after the sun has set, and under such conditions the snake may go food hunting up until the last of the daylight has faded. But in cooler weather it retires for the night soon after the sun has set, and will not reappear until sunrise the following day.

Not only are snakes disinclined to feed at low temperatures, they will also disgorge a meal they have taken during the day if the temperature falls very low during the following night. Possibly this may be due to some built-in protective mechanism which safeguards the snake from entering into hibernation with a meal in its stomach. During hibernation the snake's bodily processes are almost completely at a standstill, consequently it is unable to digest food, and its retention in the stomach undigested during hibernation could prove fatal.

Some snakes are very particular about their food being clean, and will not eat anything that has soil or other foreign matter adhering to it. On many occasions I have seen a grass snake capture a fish which has wriggled out of the snake's jaws and dropped on to loose earth. The snake will make an attempt to recapture the fish, but on finding it covered, or even partially covered, with soil it immediately releases it and will have nothing more to do with it; while if particles of earth remain in the snake's mouth as the result of picking up prey that has become dirtied, the snake appears quite distressed, and will make every effort to remove the offending matter by rubbing its open jaws against either the ground or surrounding herbage. Often this only helps to make matters worse, but the snake is not intelligent enough to realise this. Somehow or other it eventually rids itself of the pieces of earth within its mouth, possibly by swallowing them.

I have also known a grass snake to become quite disturbed by the presence of fish scales in its mouth. The snake had captured a fish which managed to struggle free, leaving some of its scales inside the snake's mouth. The fish scales were obviously causing the snake some discomfort, for it continually opened its mouth, and as it did so I caught a glimpse of the fish scales inside.

However, not all snakes, including some grass snakes, are so particular, and I have known many individuals that would swallow their prey regardless of the fact that it was almost completely covered with particles of earth or other foreign matter.

Drinking Habits

Like most land animals the snake loses moisture from its body by evaporation and this must be replaced in some way. A certain amount of fluid is probably obtained from food eaten by the snake, but during hot weather this is unlikely to be sufficient to replace that which is lost. Snakes must, therefore, drink from time to time, and the three British species are no exception. The grass snake's habitat is usually located near to water in some form or other and so it is able to drink whenever it feels so inclined. Both the adder and smooth snake, however, tend to inhabit drier localities, sometimes far removed from water in any permanent form, such as a pond or stream. But even in localities where there is no such form of water supply, moisture can at times be obtained in the form of either dew or rain, which the snake can lap with its tongue. Water can also be obtained from puddles which form, even in dry sandy areas, after a spell of heavy rain. However, all three British snakes can, when necessary, survive for quite a long time without drinking, although when given the opportunity they can do so copiously. This applies as much to the adder and the smooth snake as to the grass snake, and I have known all three species to drink without pause for several minutes. One smooth snake that I timed drank non-stop for a full five minutes, paused for about a minute, and then continued to drink for a further two and a half minutes.

When drinking a snake holds its head at an angle of about 45 degrees to the surface of the water and immerses only the front tips of its jaws. The jaws are opened only very slightly, most of the action being at the front of the jaws. Opening and closing of the jaws while drinking is performed to a steady rhythm, accompanied by a similar rhythmic contraction and expansion of the throat as the water is swallowed. That the snake does actually draw water into its mouth, and does not merely allow it to flow down its throat by gravitational pull, is illustrated by the fact that snakes can drink while hanging head downwards over the water. I have seen a grass snake drinking in this way with its head pointing directly downwards, and at right angles to the water surface, with only the tip of its nose and lower jaw immersed in the water. That the snake was actually imbibing water was evident from the rhythmic movement of its jaws and throat.

Before drinking a snake will invariably oscillate its tongue just above the surface of the water, and then touch the water with its tongue. Often a snake will drink all the water it requires in one go and without pause. But sometimes it drinks only small amounts at a time, lifting its head every ten or fifteen seconds to look around. Occasionally the head is lifted high above the water's surface, and one can observe the throat continue to pulsate for a few seconds afterwards. When pulsations in the throat have ceased the snake may lower its head to the water for another drink. It sometimes happens that a snake will yawn two or three times, or indulge in a period of tongue stretching, after taking a long drink, but this does not always occur.

In my experience only water is acceptable to British snakes. I have tried captive specimens with milk but have never known them to drink it, even though they may be thirsty. When milk has been provided the snakes have approached it with the apparent intention of drinking, but on discovering by means of the tongue that it is not water they have always moved away again without drinking. When the milk has been replaced with water the snakes have returned and drunk from it without hesitation.

Baby snakes drink equally as often as adult snakes, and I have seen a smooth snake only a month old drink without pause for seventy seconds.

That a snake can lap up water with its tongue, when the volume of water has been insufficient for it to drink in the usual manner, has been illustrated to me by the behaviour of captive snakes, which will often obtain a drink with the tongue from an all but empty water bowl. The snake does not simply flicker its tongue over the water as it would do if merely investigating, but actually pushes its tongue around the bottom of the bowl in an attempt to mop up as much water as it can.

CHAPTER SIX

Skin-shedding in snakes

The need for a change of skin – symptoms– sloughing procedure – length of shed skin – some statistics

PERIODIC shedding, or sloughing, of the outer layer of skin occurs in most vertebrates. In man it is a continuous process, the skin rubbing off in such small pieces that shedding is not noticed. Many reptiles and amphibians shed the skin in large pieces, whereas the snake usually casts its skin in one complete piece, although sometimes the skin catches on sharp or rough objects during the process of shedding and may become torn. Only in unhealthy snakes does the skin tend to flake off in pieces.

The skin that is removed is known as the stratum corneum, which is a dead, horny layer that protects the body from entry of bacteria and ultra-violet rays. It also helps to prevent excessive loss of moisture from the body by evaporation. As the stratum corneum is incapable of growth once it has been formed it must be removed at intervals to permit the snake to increase in size.

The younger the snake the more rapidly it grows, resulting in more frequent skin-shedding, and during the first year of life the skin is cast about four to five times. Thereafter skin-shedding occurs less often until maturity is reached, when the skin is shed only two to three times during the active period, which extends from about the end of March until mid-October. Skin-shedding does not necessarily occur at regular intervals, and a snake may go three months before doing so, and then cast its skin again two months later. The main factor governing the frequency of skin-shedding is temperature of the environment, as it is this that controls the snake's degree of activity, and consequently its metabolism and food intake. For this reason skin-shedding tends to occur at shorter intervals towards the end of the summer, when temperatures are generally higher, than at the beginning.

When a snake has reached the average maximum size for its species it may still continue to increase in length with each successive year of its life, but only by a very small amount. It is more usual for the snake that has reached this stage to become larger in girth rather than in length, although even this is a very slow process. However, even when body growth has almost ceased removal of the outer layer of skin is still necessary, although less often, as it becomes worn and requires replacement by a new layer.

The first indication that a snake is preparing to slough is a loss of brightness in its colours. With some snakes there is a gradual loss of colour between two to three weeks prior to sloughing, whilst in others, even of the same species, this loss of colour may not become noticeable until about ten days before the skin is shed. Gradually the snake becomes more dull and drab in appearance, and its markings less clearly defined. The underparts take on a milky appearance, and then its normally bright, clear eyes start to cloud over until they reach a stage where they are almost totally obscured by a whitish opacity. The snake is now unable to see very clearly, and possibly because of this it shows no inclination to feed. It spends most of its time basking in the sun, the warmth thus derived helping to speed up the sloughing process to some extent. The snake appears very lethargic and disinclined to move about, remaining in one place, if undisturbed, for hours at a time. Whilst unable to see clearly it is more easily taken by surprise and is, therefore, more vulnerable to attack, although its habit of remaining immobile at this time may possibly save it from being molested by other predators.

The snake's loss of colour and the opacity of its eyes are due to a thickening of the skin as the new epidermis begins to form beneath the old one. Clearing of the eyes and brightening of the snake's colours will not occur until the new outer epidermis has formed. As the new skin completes its formation the brille, or eye scale, which is part of the skin, becomes transparent like the old one covering it, thereby permitting the eye itself to be seen again. The eyes retain maximum opacity for a period of one to two days, and then gradually they begin to clear.

At the same time as the eyes begin to clear a separating fluid starts to flow between the old and the new skin. This permeates the old skin making it moist and more transparent, so that the markings and colours of the snake can once more be seen. The presence of the separating fluid is indicated by a slightly oily appearance of the snake at this stage, and also by the adhesion of small particles of grit and other foreign matter – often not apparent until the snake is examined closely – to the snake's skin.

The time that elapses between the clearing of the eyes and shedding of the skin varies considerably, and may be as little as two days, or as much as five days, although three to four days is more usual. The snake will continue to feed right up until the day before its eyes first begin to mist over. From that moment until the skin is finally shed no food is taken as a rule, although I have known some snakes to feed once the eyes have cleared.

The snake will make no attempt to remove its old skin until it is ready to be shed. When this moment arrives the snake becomes suddenly active, roaming restlessly about and rubbing its nose and sides of its head against rough objects. I have seen some snakes turn the forepart of the body sideways to the ground, and with that part of the body held stiffly, push the side of the nose along the ground. Often the snake will burrow into loose soil and push its way along just below the surface. The snake obviously feels uncomfortable about the head and jaws, as in addition to head rubbing it will constantly open its mouth, and sometimes indulge in very wide yawns. These prodigious yawns may have the desired effect of making the skin break suddenly away from either the upper or lower jaw, but not always. Quite often the skin becomes loosened around the head without actually breaking free, and as the snake exhales from the nostrils the air seeps beneath the old skin making it balloon up on top of the head. As the snake inhales the reverse procedure occurs, and the bulge disappears from the top of the head as the loosened skin collapses with an audible click against the snake's head. I have observed this behaviour more frequently in infant snakes shedding their skins for the first time, although it also occurs in the adult snake.

Eventually the skin breaks away from around the jaws and it does not take the snake long to push it back beyond the eyes by continued head rubbing. The snake seems to feel a little more comfortable once this has been achieved, and it may cease its head rubbing and bask for a while in the sun. But until its old skin is shed the snake seems unable to remain still for very long, and very soon it is on the move again, gliding through dense herbage and squeezing its way through narrow apertures, such as between plant stems, in an attempt to force the old skin back over its head and neck.

Removing the skin from the neck and the first few inches of the body is usually the most difficult part of sloughing for the snake, and may take much longer than the removal of the skin from the remainder of the body. With luck the skin may be caught and held as the snake makes its way through a narrow opening, so that all the snake has to do is to keep moving forward to pull itself out of the old skin. Sometimes it will climb into a low shrub and glide between a forked branch to achieve this end, as I have discovered from sloughs found in such places. But even without these aids the snake can shed its old skin quite rapidly once it has been pushed back along the forepart of the body. When this has been achieved a rhythmic muscular expansion and contraction of the body begins at a point just in front of where the old skin has been turned back inside out along the body. Expansion of a small section of the body at this point isolates the scales from one another, which results in the old skin being forced out from where it folds under the edge of each scale. It also pushes back the old skin a fraction of an inch along the body. The body is then contracted and expanded a little further back. This action is purely muscular and the snake makes no attempt to inflate its body with air to loosen the skin. Alternate expansion and contraction continues until the skin has been removed from about two-thirds of the snake's length. From the latter part of the body the skin usually peels away very easily, and the snake has only to glide forward and allow the drag of the slough on the surrounding vegetation to pull it free.

The snake can shed its skin by means of the muscular action described either when stationary or when moving. Its removal by this method is more rapid when the snake is on the move, and as a rule the snake combines the muscular action with forward progression through dense herbage to speed skin-shedding.

As the old skin comes away from the snake's body the presence of the separating fluid is revealed by the fact that foreign matter readily adheres to the inside of the skin being removed, and also to the snake's body from which the skin has already been removed. As the new skin dries the pieces of foreign matter are soon rubbed off as the snake moves about, but they tend to remain adhering to the shed skin. If the snake is picked up and handled while shedding its skin both the body of the snake and the loosened skin will be found to be moist and tacky.

In a normal skin-shedding the skin peels first from the head and is turned back along the snake's body, with the result that when entirely removed the skin will have been turned completely inside out. As only a very thin layer of skin has been removed it is almost completely transparent, and of a faint brown colour if cast by an adult snake. That of the infant snake is usually without colour. On the cast skins of both grass snake and smooth snake there are usually no markings visible, but on that of the adder all of the markings can be clearly seen, even though the skin is inside out. If the slough is opened up so that the outside surface can be viewed, the markings will appear almost as distinctly as on the snake itself.

The shed skin is always longer than the snake itself, often

by as much as 13 cm. This is partly due to the elasticity it possesses whilst moistened with fluid from the snake's body, which allows it to stretch a little during the process of being shed. But the chief reason for the shed skin being longer is that prior to being shed it folds under the edges of the scales. When pulled away from the scales during sloughing the underlapping skin is unfolded and extended, thus adding to both the length and girth of the slough.

As the area of skin that folds under the scales is greater over the length of the snake – there being more scales covering the snake lengthwise than around the girth – the slough is much greater in length than the snake, but only slightly bigger in girth. If the ventral surface of the slough is examined it will be found that there is approximately 2 mm of connecting skin between each of the scale markings. When totalled together this adds several inches to the length of the slough. As each of the ventral scales covers the full width of the snake's belly there is no underlapping skin tissue to allow of expansion laterally. On the dorsal surface the scale markings number more along the length of the slough than across the width, so that here also there is more connecting skin tissue lengthwise than across the width.

With the old and worn layer of skin removed and replaced by a new and much more transparent one, the snake's colours, which are in the lower layers of skin, can now be seen much more clearly, and consequently they appear brighter, as though the snake had been freshly painted. Having rid itself of the old skin the snake is once more eager to feed.

The infant snake may shed its skin within a short time of being born or up to nearly three weeks afterwards. I have seen baby adders sloughing shortly after birth, and whilst others of the same litter were still in the process of being born. Smooth snakes usually shed their skins about a week following their birth, whilst grass snakes take a little longer, sloughing for the first time between fifteen and nineteen days after hatching from the egg.

Wishing to discover the variation in size between infant snakes and their cast skins I once measured a number of baby snakes and their sloughs. In the case of the grass snake I found that whilst the length of the young snakes varied from 17 cm to 21 cm, there was not a cast skin less than 23 cm in length, whilst some measured as much as 28 cm in length. This showed that the slough of the baby grass snake could be anywhere between 6 cm and 7 cm longer than the snake itself.

Measurements of baby smooth snake sloughs revealed lengths of between 22·5 cm to 24 cm, whilst the young smooth snakes themselves varied in length from 16·5 cm to 18·5 cm. The sloughs were, therefore, between 5 cm to 6 cm longer than the snakes that shed them. The shed skins of infant adders were not measured, but most probably they would show an increase in length over that of the adders themselves similar to that found in the shed skins of infant smooth snakes.

Skin-shedding does not, of course, take place during hibernation, when the snake's bodily functions are almost at a standstill. If the snake should be preparing to slough late in the year, and the weather should happen to turn suddenly cold, causing the snake to hibernate, the sloughing process is halted, and the skin will not be cast until the following spring when the snake emerges from hibernation. Similarly, any fall in temperature during the summer when a snake is undergoing the sloughing process will delay shedding of the skin. If the temperature should drop considerably, as can sometimes happen in September or during the early spring months, and the snake become inactive, the sloughing process is suspended until the temperature rises again.

By keeping close observation on individual snakes I have found that a snake does not shed its skin at the same time each year. The snake that may, for example, have shed its skin in June and August one year, may shed it in different months the following year, and possibly with a longer interval between the two sheddings.

CHAPTER SEVEN

Basking behaviour

The need to bask – when and for how long – basking attitudes: close-coiled, flattened, etc. – sleep

IN comparison with most other wild animals, which seem to spend the greater part of their lives questing for food, the snake appears to lead a rather idle kind of existence. Needing to feed only once, or perhaps twice, in a week the snake has ample free time, most of which is spent just lying in the sun. Unlike ourselves, however, the snake has good reason for spending so much time sunbathing. For being unable to manufacture its own bodily heat, as we and many other creatures can, the snake is dependent upon external heat to provide it with the energy to perform its normal functions, such as feeding and reproduction.

At temperatures below 10° C most snakes become sluggish and inert. They show no desire to feed or mate, and this is because their physiological processes are slowed down by cold. For this reason snakes inhabiting temperate climates are forced to spend the winter months in hibernation. Conversely, the higher the temperature rises above 10° C the more active the snake becomes. There is, however, a limit to the amount of heat a snake can tolerate, and a temperature of over 45° C is said to be fatal to some species. In this country extreme cold is more of a danger to snakes than heat, for if exposed to a temperature below freezing point a snake will soon die.

In the early spring snakes only leave their hiding-places when the sun is shining. On cloudy days the temperature at this time of the year is seldom high enough to tempt them out, and even on sunny days they will only emerge when the shade temperature is above 10° C, with the temperature in the sun not less than 18° C. This is just warm enough to induce a snake to bask, but not sufficiently warm to arouse it to any degree of activity, and so it lies for hour after hour soaking up the warmth the sun provides.

Having just emerged from hibernation, during which time its bodily functions have been practically at a standstill, the snake requires a week or more of almost continuous sunshine to reactivate its sluggish metabolism before it can even begin to feel the desire to mate, and it is not until the shade temperature reaches 15° C, or about 21° to 24° C in the sun, that the snake starts to become really active. At this temperature it is sufficiently alert and quick enough in movement to to be capable of capturing its prey, although in the spring it shows little desire for food, and its re-awakened energy is directed towards courtship and mating.

When the mating season comes to an end around the middle or end of May, and the snake's need for food replaces its desire to copulate, the temperature on cloudy days is frequently below that required to induce a snake to feed, and so the snake must wait for the sun to shine before it can take a meal. As the summer advances and daytime temperatures steadily rise the snake needs to spend less time basking in the sun to attain the degree of activity required to enable it to hunt down and capture its prey. And so whereas in early summer a snake may require to bask for an hour or two before it feels the desire to feed, less time is spent basking as the weather becomes warmer. In middle and late summer the temperature at daybreak is often high enough for the snake to go food hunting without first needing to warm itself in the sun, although after having fed it may spend an hour or so basking before the sun's heat becomes too intense. For warmth is also required by the snake to enable it to digest its meal, and the warmer the snake is the quicker its meal is digested. Speedy digestion is of prime importance to the snake, as food swallowed whole creates a weighty bulge in the body, which impedes rapid movement, making the snake more vulnerable to attack. Even when the snake is able to bask for long periods after swallowing a large meal, digestion may take three to four days, and this time may be extended to a week or longer if the snake is unable to bask, and the temperature of its surroundings remains low.

To enable it to obtain and conserve the maximum amount of warmth from the sun the snake has developed different basking attitudes to accommodate prevailing circumstances. In the early spring when the sun's power is not very great, even at midday, the snake adopts a close-coiled position, which it achieves by moving round and round on the spot until it is coiled up as tightly and as neatly as a Catherine-wheel. Sometimes, instead of moving round in circles, the snake travels back and forth to attain an elongated pattern of coils which lie close together, with the snake's head resting on its coiled body at one end of the pattern. These close-coiled attitudes are also used in the early morning or late afternoon during summer when the sun is low in the sky. By basking in this way, with the coils of the body touching, the snake is better able to conserve the warmth it receives from the sun, which would be more quickly dissipated if the snake were to lie more loosely coiled. Once the snake has become thoroughly warmed by the sun it may still continue to bask but in a more loosely coiled attitude.

A habit peculiar to the adder is that of flattening the body while basking. Not all adders adopt this attitude when sun-basking, and those that do employ it mainly when they first emerge to bask, although they may bask continuously in this manner if the sun is weak. The purpose of this behaviour is undoubtedly to catch as much as possible of the sun's rays, as by flattening the body no one side is in shadow, and the largest possible area is presented to the sun. This attitude would be particularly useful when the sun is low in the sky, for example, in the early morning when the adder first emerges to bask, and requires to warm itself rapidly. If the sun shines continuously it takes the adder about thirty minutes to become thoroughly warmed through, and once this has been achieved the adder reverts to its normal cylindrical shape.

I have never observed either grass snake or smooth snake flatten the body when basking, and with the adder it appears to be achieved by extending the ribs outwards so that they are more or less straight, instead of in their usual curved position. The body is by this method flattened throughout its length, and appears to be about one quarter of an inch in depth and twice its normal width, while the spots along the sides of the body seem to have moved on to the snake's back.

Thomas Bell mentions in his *History of British Reptiles* (1839) the adder's habit of flattening its body when basking. He states, '. . . during the period of gestation it is very common to see the pregnant female Viper lying, as it were, all abroad, with flattened body, basking in the sunshine'.

Having studied closely the basking behaviour of snakes over a considerable number of years I have come to the conclusion that although the snake is unable to close its eyes (having no eyelids), it does at times become unconscious whilst sun-basking. When actually asleep a snake will often allow its head to hang loosely over its coiled body, with the lower jaw resting on the ground, almost as though the reptile were dead. It will sometimes remain in this position for an hour or more, and appears not to see movements made quite close to it. Even vibrations of the surface upon which it is resting do not disturb it when it is in this unconscious state, whereas when basking in the ordinary way, the slightest vibration or movement near to it would cause it to lift its head enquiringly. If the vibrations are repeated the sleeping snake will, however, awaken with a start and be instantly on the alert, ready to vanish quickly into hiding should it feel insecure.

Once whilst watching several captive adders basking together I saw one, that was basking in the sleeping position just described, dislodged from its place on a log by the movements of another snake. The basking adder slipped and fell, still coiled, on to its back, but instead of righting itself immediately, it continued to lie upside down for several minutes. Then it slowly turned over and made its way to another spot to continue its basking. As the adder in question was a plump, well-fed specimen that had been captive only a few days there was no question of it being sick or weak, as might be suggested by its apparent lassitude.

It is not unusual to find a snake that is about to slough basking in the sun, for warmth hastens sloughing. Consequently a snake will spend as much time as it can basking prior to sloughing, often lying unmoving for hours in one spot until the time for casting the skin arrives.

During the early spring snakes may be found basking at almost any time of the day, but during the warmer months of the year basking is mostly done during the morning. There can be no hard and fast rules about this, however, as much depends on prevailing weather conditions. For example, after a long spell of cloudy weather snakes will emerge to bask the moment the sun shows itself, regardless of what time of the day this occurs. Similarly, if the sunshine is intermittent throughout the day snakes will come out to bask at almost any time from sunrise to sunset. But during settled spells of warm, sunny weather early morning basking is the general rule. From the beginning of May until the middle of September a cloudless sky will bring the shade temperature up to around 21° C by about 11 a.m. on a calm day. When the shade temperature rises above this degree most snakes disappear under cover, for although they love warmth they cannot stand intense heat. Of the three British species the adder is the most averse to great heat, and when the temperature starts to rise above 21° C in the shade it is invariably the first to seek shelter from the sun. Indeed, the adder's liking for moderate temperatures is the reason why it is found further north, both in Britain and in Europe, than either grass snake or smooth snake. The smooth snake seems the least affected by heat, and will continue to lie in the sun for some time after the adder and grass snake have vacated their basking places.

CHAPTER EIGHT

Hibernation

Duration – correlation between temperature and activity – hibernacula – mild winters

BEING cold-blooded and therefore unable to remain active at low temperatures, the snake inhabiting a temperate climate is forced to spend the colder months of the year in hibernation.

In Britain hibernation for snakes often begins about the end of September, but may be delayed until the middle, or even the end, of October if the weather is warm and sunny. I have found adders out basking on 5 October, and grass snakes as late as 10 October, in the south-east of England, but in a warm October both grass snakes and adders have been seen basking in some parts of the country until the end of the month. On average hibernation lasts approximately five months, but may sometimes be shorter when spring arrives early.

Of the three British species the adder is the most resistant to cold, and can remain active at a temperature as low as 8° C. Consequently, the adder's period of hibernation tends to be a little shorter than that of the grass snake and the smooth snake. All three species, however, are disinclined to feed when the temperature falls below 15° C, and they spend the warmer days of autumn just lying in the sun.

In parts of North America where winter temperatures may drop as low as minus 20° C, snakes require an hibernaculum deep underground where they will be safe from frost. Usually a deep crevice in the side of a mountain is chosen, and as such places are not everywhere to be found it is not unusual for scores, and sometimes hundreds, of snakes, often of different species, to share the same winter hiding-place. To reach the winter den a snake may have to travel some distance, so that when late summer nights start to get cooler it begins its slow migration away from its summer haunts towards the communal hibernaculum.

In Britain very low air temperatures are only rarely experienced, and when they occur they are insufficiently prolonged to result in the soil temperature becoming very low. In consequence it is not necessary for British snakes to retire very far below ground to escape the frost. Because of this they can hibernate almost anywhere, and do not need to leave their summer haunts to seek a special place in which to spend the winter. Communal hibernacula containing dozens of snakes entwined together have been discovered in this country, but communal hibernation is the exception rather than the rule, and most snakes hibernate singly. Many spend the winter buried in the loose soil around the roots of a bush, or among heather roots. Snakes do not make their own burrows, but can push their way into loose soil, although often as not they make use of the tunnels made by mice to get below ground. Disused rabbit burrows are also used by snakes as places in which to hibernate, while some spend the winter buried in the loose soil beneath piles of dead wood and bracken. It is unusual to find snakes hibernating in places from which they could easily be disturbed, such as beneath dead logs or large stones.

Hibernating grass snakes and adders have been discovered at depths varying between 25 and 122 cm. Those found at shallower depths were usually among heather roots or the roots of small bushes, whilst those found at the deeper levels had made use of disused burrows made by other animals in dry banks.

In order to discover for myself just how deep into the earth a snake would have to descend to escape temperatures below freezing point I conducted an experiment in the winter and early spring of 1969. I placed thermometers at depths of 25, 50, and 100 cm, and kept a record of the readings obtained from the beginning of February until the end of

April. Up until the end of January that year the weather had been quite mild, and it was not until the beginning of February that temperatures began to fall. Charts showing the temperatures that were recorded below ground, in comparison with the minimum and maximum air temperatures, are to be found on page 97.

One thing that I discovered, and which surprised me considerably, was that at a depth of only 25 cm below ground the temperature never fell below 2° C, even when the air temperature dropped as low as minus 7° C. At 50 cm below ground the lowest temperature reached was 4·5° C, whilst at 100 cm it was never below 5·5° C. At a depth of 25 cm the fluctuation in temperature from the beginning of February to the end of March (when hibernation has usually ended) was never more than 5 degrees, although the air temperature had fluctuated by as much as 20 degrees. At 50 cm the temperature had remained even more constant and varied by only 3·5 degrees during the same period. From 9 February to 5 April the variation in temperature at 50 cm was as little as 1·5 degrees; whilst at a depth of 100 cm the variation in temperature from 2 February to 31 March had amounted to 2·5 degrees, and to only half a degree from 13 February to 31 March.

Spring was late arriving in 1969 and the daytime air temperature did not rise above 12·5° C until the last day of March. Up until that date the variation between the day and the night temperatures at a depth of 25 cm was no more than half a degree. When, however, the air temperature rose suddenly during the day on 7 April to a maximum of 22° C, and continued to rise over the next three days to a maximum of 23° C, there was a variation between the day and the night temperatures of between one and 3·5 degrees at that depth. At the 50 cm level there was no variation over the same period between the day and the night temperature, which rose steadily from 6° to 8° C. When the air temperature had reached 23° C the temperature at 100 cm below ground was still only 7° C, which was just one degree higher than it had been throughout the coldest part of the winter.

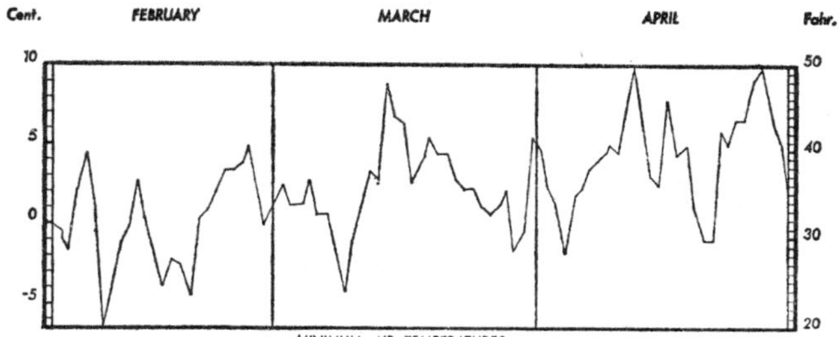

MINIMUM AIR TEMPERATURES
(Recorded 1969)

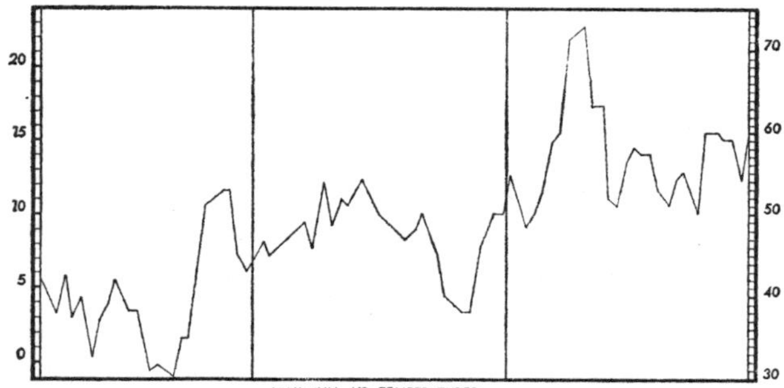

MAXIMUM AIR TEMPERATURES
(Recorded 1969)

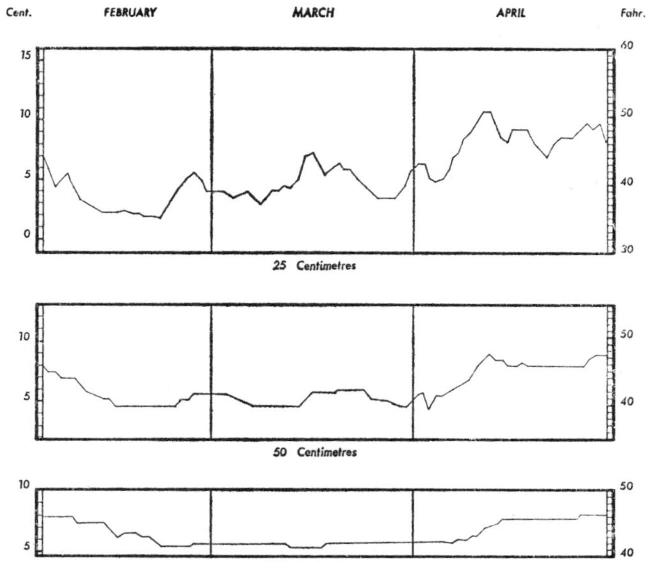

SOIL TEMPERATURES AT DIFFERENT DEPTHS
(Recorded 1969)

This simple experiment proved that at a depth of only 25 cm below ground a hibernating snake would be safe from frost during an average winter, although in a very severe winter, where a prolonged spell of low temperatures could occur, the frost might just reach the snake. At a depth of 50 cm in the soil a snake could pass through all but the most exceptionally severe winter conditions, whilst at depths beyond 50 cm it could survive the very worst that an English winter could produce.

The fact that at depths below ground of 25 cm and beyond the temperature responds only very slowly to changes in air temperature explains how snakes are able to hibernate undisturbed by the sudden changes of weather that we experience in this country. At depths just beyond 25 cm the body temperature of a hibernating snake would never vary more than about 4·5 degrees from the start of the winter until the beginning of spring, and this constant cool temperature would enable the snake to maintain its condition, with only slight loss of body-weight, until rising temperatures brought it out of hibernation. Furthermore, the delayed response of the soil temperature in relation to rising air temperatures would ensure that the snake did not emerge until the really cold weather was over. However, prolonged spells of unusually mild weather during the winter will sometimes bring snakes out of hibernation before the proper time. The adder has been seen basking in the sun as early as February, and sometimes when there has been snow lying on the ground. This might appear to indicate that this snake will emerge from hibernation at very low temperatures, but such is not really the case.

A prolonged period of very mild weather in mid-winter could raise the temperature of the adder's hibernaculum below ground to 8° C, which is the minimum temperature required to rouse the adder from hibernation. A subsequent sharp fall in air temperature, to a degree sufficiently low to induce precipitation in the form of snow, would not result in an immediate lowering of subsoil temperatures, as can be seen from the chart on page 97. Consequently, the adder might emerge from its winter hiding-place unaware that conditions above ground had become much colder than those below. On emerging the adder might find the sun shining warmly enough to permit it to bask, even though the shade temperature is likely to be lower than that required to make a snake active.

In an average winter the temperature of the soil at a depth of about 25 cm does not reach 8° C until the air temperature is around 13° C in the shade, which is not usually attained until some time in March. Following a very cold winter the subsoil temperature would possibly respond more slowly to a rise in air temperature and may, perhaps, not reach 8° C until the air temperature had risen some way above 13° C.

The earliest I have found adders out basking was 3 March, when the air temperature reached a maximum of 14·5° C, which, as can be seen from the temperature charts, would give a subsoil temperature at a depth of 25 cm of about 8° C. As mentioned earlier, this temperature is just sufficiently high enough to rouse the adder from hibernation.

Assuming a snake were to be drawn from its hibernaculum in mid-winter by exceptionally mild weather conditions, it is unlikely to be killed by a sudden frost, for as soon as the air temperature dropped below 8° C it would retire to its hibernaculum once more.

As the snake is capable of surviving for long periods without food when active, hibernating for several months without feeding presents no problem. At low temperatures the snake's metabolism is slowed to such an extent that it does not require food, and it emerges from hibernation looking as plump and well nourished as when hibernation began. It will have used very little of the reserves of fat which it carries, and which help it to survive during the first few weeks after hibernation has ended, when it is too preoccupied with courtship and mating to have much time to spare for food hunting.

CHAPTER NINE

How the adder bites

The fangs – the venom – hunting – self-defence – the strike – handling adders – adder v. *hedgehog*

THE viper family, of which the British adder is a member, possess the most advanced poison fang development found among snakes. The fangs themselves are two enlarged teeth, recurved in shape, situated at the front of the upper jaw. They are not completely hollow, the centre being filled with pulp as in any ordinary tooth, but at the front of the fang a fine tube has been formed by a folding over of the dentine, or bony structure, of the tooth, and it is through this that the venom flows, emerging through a tiny aperture at the front, close to the tip of the fang.

Whereas many species of poisonous snake have poison fangs which are comparatively short and rigidly fixed in the mouth, all those species belonging to the viper family possess fangs of such great length that it would be difficult for them to close the mouth if they were rigidly attached to the jawbone. This problem has been overcome during the slow process of evolution by a modification of the jaw bones which permits the poison fangs to be folded back along the roof of the mouth. This is made possible by the fact that each of the poison fangs is attached, independently of the other, to the base or posterior end of a small piece of bone known as the maxilla, which hinges loosely on the prefrontal bone at the anterior end of the skull. The maxilla is also hinged to the pterygoid bone which carries the smaller teeth in the snake's upper jaw. Erection of the poison fangs into the vertical position is accompanied by a forward movement of the paired pterygoid, which in turn push the two maxillae forward, thereby bringing the poison fangs from their recumbent position into a vertical position where they are ready for use.

Each of the two fangs is supplied by a separate venom gland, situated just below and a little to the rear of the eye. When the adder bites, venom is squeezed by a muscular action from the gland and along a narrow duct or tube to the base of the fang. The duct does not enter the fang but is connected to the gum which folds around the base of it. The venom flows from the duct into the pocket formed by the gum, and from there it enters the fang through a small aperture at its base. This arrangement permits the poison conducting fangs to be shed periodically and replaced by one of the reserve fangs, which then becomes immediately operational. If the poison duct were directly attached to the base of the poison fang, the connection between the two would be ruptured when the fang was shed, and the snake would be unable to use the reserve fang until the duct had re-attached itself to the base of it.

Although there is a cluster of reserve fangs in various stages of growth, only the leading fang receives venom. When that fang becomes detached – almost invariably being imbedded into a meal and swallowed along with it – the reserve fang that was immediately behind it is already in position to accommodate the flow of venom from the gland. Shedding of the poison fangs takes place every few weeks, although both fangs are not necessarily shed at the same time. The leading fang on either side of the jaw is pushed from its position at the base of the maxilla bone by the gradual moving forward of the reserve fang immediately behind it, which, in turn, is replaced by the reserve fang next in line as it reaches full size.

The poison conducting fangs of an adult adder measure approximately 7 mm in length, and each fang, together with its cluster of reserve fangs, is covered by a fleshy shroud to protect it from possible damage, such as might occur when bulky prey is being consumed. This shroud or sheath completely conceals the poison fangs when they are in the recumbent position, and almost completely so when they are erected, only the fine tips of the fangs being visible. The poison fangs are, therefore, protected at all times by their fleshy shroud, which is pushed back only when the fangs enter the flesh of a victim, although the fangs are partially revealed when the adder opens its jaws particularly wide, as for example, when yawning, and when lunging forward to strike.

While it is true to say that because of the length of the poison fangs it would be difficult for the adder to close its jaws with the fangs erected, it is not altogether impossible, for I have seen an adder close its jaws with the poison fangs still erected and hanging down over the lower jaw in a manner reminiscent of the prehistoric sabre-toothed tiger. Such behaviour in the adder is, of course, unusual, as almost invariably the fangs are returned immediately to the recumbent position after use. In the instance quoted, however, the adder had made repeated use of its fangs, and was probably too tired to return them to their normal position. It does, however, serve to illustrate that the fangs do not automatically fold into the recumbent position when the jaws are closed. Conversely, the fangs can remain folded when the jaws are opened wide as, for example, when large prey is being swallowed. Nevertheless, it is usual for the fangs to be erected whenever the jaws are opened wide, and apart from when feeding this only happens when the snake is either about to strike, or when it yawns, which it does frequently. I have never yet seen an adder yawn without erecting its poison fangs, and the folding of the fangs is usually synonymous with the closing of the jaws.

Adder venom is a pale amber in colour and the average amount injected by an adult adder at one bite is somewhere in the region of 0·01 cc. It is a highly complex substance composed largely of various proteins and enzymes. In a dehydrated condition snake venom has been known to retain its potency for a period of up to forty years.

The chief purpose of the adder's venom is that of overpowering prey, which it does very rapidly, a single bite from an adder being sufficient to kill a common lizard within two minutes. A secondary use is that of defence, although some adders seem loath to use it for this purpose, and will only bite on extreme provocation. The bite of an adder is really more in the nature of a stab, and is performed at very high speed. When about to deliver a poisonous bite the adder usually draws up the forepart of its body into an S-shaped loop. By quickly unfolding this loop the adder can extend its head forward in a lunging strike at its victim. As the adder's head travels forward the jaws are opened wide and the poison fangs are erected. In a fraction of a second the adder stabs its poison fangs into its victim, withdraws them with equal speed, and recoils itself in readiness to strike again if necessary. The whole action of striking and returning to the coiled position is executed so rapidly that it is almost too fast for the eye to follow. Just how fast the action is can be gathered from the fact that when I attempted to photograph an adder striking I found that an exposure of 1/2000th of a second resulted in a blurred image on the film, and that blurring still occurred at times when the exposure was reduced to 1/3000th of a second.

When hunting its prey the adder usually approaches it by stealth until within striking distance. It then adopts the attitude just described and waits for its intended victim to make a move, striking the moment it does so. Although normally very accurate in its aim, the adder sometimes misses the target, and its intended victim makes a dash for cover. When this happens the adder will sometimes wait for a moment and then go in search of its prey. Equally as often the adder gives chase, moving with surprising speed, and strikes at its prey whilst on the move. I have seen an adder strike in this way while in pursuit of a fleeing lizard. On these occasions it has no time to prepare itself for the strike, but simply propels itself forward a little faster to catch up with the lizard and drives its fangs home. At other times the adder will make an unprepared, but often effective, strike at its prey when it doubles back whilst being pursued.

When defending itself against an aggressor the adder invariably coils itself compactly so that it is less vulnerable to attack. The forepart of the body is folded laterally in an S-shaped loop, while the nose of the reptile points in the direction from which the attack is expected. Where this is above the adder's own level, for example when the aggressor is either a large animal or a human being, the adder will sometimes lift the forepart of its body at an angle of as much as 45 degrees from the ground.

Individual adders vary somewhat in their reaction to attack, and while a few remain quite calm and controlled in the face of danger, most become excited and hiss loudly with both inhalation and exhalation of air from the body, the rise and fall of the body as it is alternately inflated and deflated giving the impression of intense anger. In reality the adder is a very timid and retiring creature, not given to making unprovoked attacks upon creatures too large for it to feed upon. It will always choose to retreat from an intruder, if it can find a way of escape, rather than stay to fight, but if cornered, or caught out in the open away from cover, it will defend itself viciously.

The adder's maximum striking range is about 13 to 15 cm, and the direction of its strike can be altered in a split second to cope with any last moment change of tactics its attacker may make. If a pair of gloved hands are extended from opposite directions towards an adder on the defensive, the creature's head will swivel with amazing rapidity as it turns to face first one and then the other of the approaching hands, and it will always strike first at the one that is nearest.

Should one strike prove ineffective in deterring its aggressor the adder will strike again and again, its first few strikes being the most accurate, and the speed with which they are delivered making them difficult to avoid. But the adder tires quickly, and after a number of strikes made in quick succession its aim becomes less accurate, and the speed of each successive strike a little slower, the target occasionally being missed altogether. It does not take the adder long to learn that its venom is ineffective against inanimate objects that may be thrust towards it, and it will eventually become reluctant to strike at such objects. It seems puzzled by the ineffectiveness of its venom and will often extend its tongue towards the object it has been striking, as if trying to determine its nature. Having satisfied itself that the object is not a living thing, it may thereafter either refuse to strike at it or strike with the mouth closed, such strikes sometimes being delivered with as much force as when the fangs are used, and the impact is considerable in relation to the reptile's small size.

In my experience the adder will not strike at a stick thrust towards it. But if the end of the stick is made to resemble a human finger by padding it and covering it with either a piece of cloth or a finger cut from a glove, then the adder will strike at it. Having discovered this I later developed the idea to obtain a photograph of an adder striking. My first attempts were unsuccessful because I operated the camera manually when the adder struck at the imitation finger. I soon found that the adder's strike was so fast that by the time I had pressed the shutter release the adder had already struck and recoiled. I realised that the only way I was likely to get the picture I wanted was to let the snake take its own photograph. This meant that the adder would have to operate the camera by a remote release, and to make this possible I attached to the end of a short stick a camera triggering device of my own construction, which I padded with foam plastic, to prevent damage to the adder's fangs, and concealed with a finger cut from a soft leather glove.

This device worked admirably, although some adders struck more readily at it than others. I found that where an adder showed reluctance to strike at this imitation human finger it could be induced to do so if I transferred my own scent to it. This was done simply by holding the device in my hands for a while, before offering it to the adder to bite at. I found, however, that most adders would strike more readily at a hand, even when the hand was covered with a glove. Nevertheless, I was surprised by the lack of aggressiveness exhibited by most of the adders I experimented with. Occasionally I came across an adder that would strike repeatedly, but the majority would strike only once or twice and then refuse to co-operate. Some would make a couple of strikes using their fangs, and thereafter make only feinting strikes at the object thrust towards them, not actually making contact with it. Where an adder could be induced to strike repeatedly, the first few strikes were found to be delivered with the greatest force, and with each strike venom was emitted. After half a dozen or so strikes very little or no venom was emitted, which seemed to indicate that the venom sacs had been exhausted. I found that most adders struck from the close-coiled position, although a few would strike whilst lying more extended. Occasionally an adder would strike with such force that it lifted itself bodily about an inch off the ground, this most usually happening when the adder was lying in an extended position.

The close-coiled defensive posture which I have already described is only adopted by the adder when it is cornered, or when it feels it is unable to escape without being attacked whilst doing so. Should the adder be attacked from the rear whilst in the act of fleeing from its aggressor no attempt at a defensive pose is made, the adder simply turning very quickly and plunging its poison fangs into whatever part of its aggressor happens to be nearest to it.

When I first began studying the adder I never wore gloves to capture specimens I wished to take home. My method was simply to lay a short stick across the back of the adder, as near to the head as possible, grasp the snake by the tip of the tail and lift it quickly from the ground. This is a fairly safe way to handle an adder, since, being rather thick-bodied, it has difficulty in swinging itself up to bite the hand that holds it. One needs to be careful, however, that the snake does not climb up its own length and reach one's hand in that way. Usually a shake of the hand is sufficient to prevent it from doing this, but one needs to take care, and to be prepared to drop the snake in a hurry. As a result of this method of picking up adders I was eventually bitten by one, and prior to that I narrowly escaped being bitten on several occasions. Familiarity in handling these reptiles resulted in my becoming a little careless at times when trying to capture them, and on more than one occasion I have caught hold of an adder by the tail just as it was disappearing into the herbage. Being unable to see the front half of the adder under those circumstances it was impossible for me to lay a stick across its back, and the adder would hitch the front portion of its body around the stems of plants to prevent itself being pulled from the herbage by the tail. These tug-of-wars were frequently terminated by the adder turning and striking at my hand. As the adder's head would be in amongst the herbage it could not see at what it was striking, but simply aimed in the direction of its own tail, which I was holding. Consequently the adder either missed its target, or I was able to let go in time. On one occasion an adder's fangs just skimmed the top of my thumb. I immediately released my hold and the adder quickly vanished from sight.

When first caught and held suspended by the tail the adder, quite naturally, is very alarmed. It usually hisses loudly and strikes wildly in all directions. It appears to be in an awful rage, but more probably is very frightened. Although there may be nothing within its reach it continues to strike at the empty air, and all the while it squirms and wriggles frantically to escape. Not all adders react to capture in this way, and I have caught immature specimens which made no attempt at all to strike. Male adders usually struggle more vigorously to escape than do females, but this is not always the case. I have on occasion picked up a male adder that has remained quite calm and made no attempt to bite. The male is, however, slimmer in build than the female and is, therefore, more energetic as a rule in its efforts to free itself. Large females are often quite sluggish, particularly when they are carrying young, and many put up little resistance once caught. The adder is invariably more intractable in hot weather when it is more alert and active. At low temperatures it is much easier to handle.

When I first began studying British snakes I was in the habit of frequently handling those which I caught and brought home. But I now make a point of handling them as little as possible while they are in captivity so that they do not become tame. For the same reason I endeavour to keep out of their sight until the time comes for them to be returned to their natural habitat. In those early years, however, I often tried to tame adders, and I found that they were not the vicious untameable creatures that some people would have one believe.

After an adder had been captive for a few days I would lift it gently on the end of a stick from its cage, and then transfer it slowly to my gloved hands. Usually the adder would make no attempt to strike, and I would allow it to glide slowly from one hand to the other. Every movement had to be performed slowly and deliberately, with no sudden movements made that might alarm the adder. When handled in this way the adder invariably remained calm and would not even hiss. If it did become alarmed and began trying to escape I would allow it to glide down on to the ground, and there let it tire itself out before attempting to pick it up again. If, instead of doing this, I tried to restrain the snake from escaping from my hands it would immediately turn and bite. Usually this was performed with a rapid strike, but not always. I have known an adder to behave quite calmly on being handled and then, for no apparent reason, slowly and deliberately bite my gloved hand, chewing its fangs into the thick leather as if determined to reach the flesh beneath. On one occasion whilst handling an adder it turned suddenly and struck at my hands, and to my surprise it missed and plunged its poison fangs into its own body. Whether or not venom was injected I do not know, but certainly that adder suffered no ill effects from the bite.

But as a rule the adder would make no attempt to bite if handled gently, and after several sessions of being handled with gloved hands it would appear to have become quite tame. I would then remove the gloves from my hands and allow the adder to glide through them as before. I invariably chose cool days for these snake-charming acts of mine, as snakes are less excitable then than they are in hot weather. When I first handled an adder with my bare hands I fully expected to be bitten the moment it picked up the human scent from my hands. But obviously the adder had become accustomed to my scent as the result of close contact before the gloves were removed. Also, having learned that it had nothing to fear, and knowing that it would be gently treated, it did not become alarmed on being picked up.

When I became accustomed to handling adders with my bare hands I decided one day to try something I had read about in a book called *The Quiet World of Nature*. The author, Bernard Gooch, related therein how he allowed the wild adders he found in his garden to climb up his arms and from there on to his head. I thought I would like to try it, and so in the quietness of my room I lifted an adder from its cage and let it climb up my arm, on to my shoulder, and from there on to my head. I kept as still as I could, and I must confess to feeling a little apprehensive as the adder explored my hair, and it occurred to me that there was every likelihood of its poison fangs entering a major blood vessel if it should choose to bite at that moment. But I dared not move my hands to lift it down in case it should become alarmed, and so I had to wait patiently until the creature descended of its own accord. I realised afterwards that it was a rather foolish experiment to try, and I never repeated it. It did, however, prove to me quite conclusively that the adder is not, as one writer put it, 'as untameable as the Scottish wild cat'.

Some years after trying the experiment just described I discovered that Bernard Gooch did not, in fact, allow freshly caught adders to explore his person, and that he first got them accustomed to being handled, as I myself had done.

That some animals are immune to the effects of snake venom is a well-known fact. Most poisonous snakes are in no way affected when their own venom is injected into them, whilst some are also known to be immune to the venom of poisonous snakes belonging to a different species, the venom of which may be of a completely different type. Even more surprising is the fact that many harmless snakes are immune to the bites of poisonous snakes, an example of which being the king snake of North America which will attack and devour without hesitation the venomous rattlesnake.

It has been said that the hedgehog is immune to the bite

of the adder, but there is no concrete evidence of this to my knowledge. The idea that the hedgehog possesses such immunity has stemmed from the fact that the hedgehog will attack and eat the adder when it gets the opportunity. This does not, however, prove that the hedgehog is unaffected by the bite of an adder, as the hedgehog is well protected by its spines, the only parts of its body that are vulnerable being the face and legs.

To discover what would happen when hedgehog and adder met I once placed the two together and awaited events. The adder took little notice of the hedgehog at first until the latter shuffled up to it and promptly bit it on the tail. The adder reacted with a loud hiss, whereupon the hedgehog immediately tucked in its head. About thirty seconds later the hedgehog lifted its head and again bit the adder on the tail. Another loud hiss from the adder was followed by a strike, the snake's open jaws hitting the spines covering the back of the hedgehog's head, which the creature had cunningly tucked in again. Having struck once the adder did not seem inclined to repeat the act, having found, no doubt, the points of the hedgehog's spines rather painful to its mouth. The hedgehog was persistent though, and once more bit the adder in the same place. At this point the adder seemed to decide that he was no match for his spiny aggressor and began to move off. But the hedgehog shuffled after it, repeatedly nipping it on the tail and then tucking in its head. The adder was growing more and more angry, or perhaps alarmed, for it now kept up a continuous hissing, which seemed to deter the hedgehog, for it remained with its head tucked beneath its body, although it was not completely curled up, for much longer. Meanwhile the adder, having found that it was unable to move away from the hedgehog without getting bitten in the tail, was now coiled in a defensive attitude, its body alternately expanding and deflating as it continued to hiss unceasingly. It dared not strike at the hedgehog for fear of injuring itself on the creature's spines, but its hissing at least prevented the hedgehog from biting. As the hedgehog made no attempt to uncurl the adder eventually stopped hissing and made towards it, and to my astonishment it climbed on to the hedgehog's back, where it settled itself down as though to bask. Doubtless the adder was attracted by the warmth of the hedgehog's body, as it was a somewhat cool day. But after lying coiled on top of the hedgehog for only a few minutes the adder appeared to find the sharp spines a little uncomfortable and descended from its lofty perch. All the while the adder was on the hedgehog's back the latter made no attempt either to move or lift its head. Obviously it had no intention of exposing itself to danger whilst the adder was in such close proximity.

At this stage I separated the two creatures, and on examining the adder I could find no mark of any kind to show where the hedgehog had bitten it. This was most surprising as the hedgehog had appeared to be biting the adder quite hard. That the hedgehog was not afraid of the adder was obvious from its persistent attacks on the reptile. But the fact that the hedgehog adopted a defensive attitude whenever the adder hissed, and also immediately after each bite it took at the adder, seemed to indicate that it possessed no immunity to adder venom, and that it was aware of the risks attendant upon attacking the reptile.

CHAPTER TEN

The effects of adder-bite

The action of adder venom – size of injection – symptoms – first-hand experience – case histories: people and dogs

WHILE each species of venomous snake has its own particular brand of venom, which differs in varying degrees from that of any other, most snake venoms are either predominantly haemotoxic or neurotoxic in their action. The former destroys the red blood cells and also causes haemorrhage of the blood vessels, while the latter attacks the nerve centres, which control the automatic function of both heart and lungs.

The venom of the cobras, mamba, and sea snakes is of the neurotoxic type, while that of the vipers and rattlesnakes is haemotoxic. However, few snake venoms are purely neurotoxic or entirely haemotoxic in action, and many are a mixture of the two types, one of which predominates. For example, the venoms of the spectacled cobra, Australian black snake, and the Australian copperhead snake, although chiefly neurotoxic, also contains a blood destroying agent. Conversely, the mainly haemotoxic venoms of the timber rattlesnake, puff adder, and carpet viper contain a neurotoxic element.

The venom of the British adder is chiefly haemotoxic. It destroys the red blood cells, causes haemorrhage of the blood vessels, and can produce intravascular thrombosis which impedes blood circulation. In addition it also has a depressive action on the heart.

Research has revealed that venom stored for some time in the venom glands of a snake is of a higher specific gravity, and therefore more toxic, than venom recently manufactured by the venom glands. Consequently the bite of one adder can sometimes prove more toxic than that of another, even though the same quantity of venom may have been injected. In other words a bite from an adder that has not used its poison fangs for some while is likely to result in more severe symptoms than a bite from an adder that has recently used its fangs, either for feeding or defence purposes. There could, therefore, be some truth in the old country belief that the bite of an adder is more dangerous in the spring than at other times of the year. For mating takes place in the spring and during that time the adder shows little interest in food. Venom would, however, be manufactured by the venom glands and stored for future use.

The amount of venom an adder injects with a single bite is sufficient to bring about within minutes the death of the small animals upon which it feeds. But the same quantity of venom injected into a larger creature, such as man, would not normally constitute a lethal dose, although it can often produce very alarming symptoms.

The severity in humans of the symptoms resulting from adder-bite is dependent upon a number of different factors, including the size of the adder, the age and state of health of the victim, and also the amount of venom injected. The latter can vary considerably, and is not necessarily governed by the size of the adder. A large adder could, of course, inject a larger dose of venom than a small one, but this does not always happen. For example, if the muscles which squeeze the venom from the venom glands are not fully contracted when the bite is delivered a sub-average dose may be injected. It quite frequently happens that only one of the two poison fangs actually pierces the victim's flesh, in which case only half the normal dose is injected. But even when both fangs enter the flesh, venom may be expelled from only one of them. This can happen when an adder has recently shed one of its poison fangs before the reserve fang is properly in position to receive venom from the gland. Sometimes clothing prevents the poison fangs from being fully imbedded into the victim, so that only some of the venom is absorbed into the bloodstream. The same thing occurs when a bite is received on a part of the body well protected by fleshy tissue, the venom being absorbed by the tissues and thereby inhibited from rapid absorption into the main bloodstream. This helps to localise the effects of the bite, and gives the body time to muster its defences against the toxin. Going to the other extreme, a bite where the fangs actually pierce a major blood vessel would allow the body little time for defensive action. The venom would quickly circulate to the heart where its depressive action, in addition to the damage caused by haemorrhage and coagulation of the blood cells, would be severe enough to present a real danger to the victim's life. Fortunately this rarely happens, most bites occurring on the fleshy parts of the extremities.

A survey taken some years ago revealed that the majority of adder-bites are received on either the ankles or the hands, more usually the latter. A bite on the ankle would most probably result from accidentally stepping on or near a basking adder, whereas a bite on the hand is, more often than not, the result of indiscretion on the part of the victim. A few unfortunate people do occasionally get bitten on the hand whilst gathering wild flowers, or when climbing banks and slopes, but judging from the numerous reports of adder-bite I have had sent to me, it appears that in the main a bite on the hand is the direct consequence of attempting the capture of an adder.

A single bite from an adder rarely proves fatal to an adult person, as the amount of venom injected is small in relation to the size of the victim. Accordingly young children, because of their smaller size, are more seriously affected as a rule, although seldom fatally. Excepting those in bad health, it is the very young and the elderly who are generally the worst affected. In the very young the symptoms may be more severe in the initial stages, but recovery is usually rapid, whereas in elderly people the symptoms are normally less serious, but recovery is much slower. Fear of the consequences can, and sometimes does, cause the snake-bite victim as much harm as the bite itself. For this reason it is important that the adder-bite victim should be firmly reassured that he is not going to die.

Being bitten by an adder is rather like being stabbed very hard with a needle, or, to be more exact, two needles. Apart from the initial stab of the fangs into the flesh no pain is felt at first, but within ten minutes or so of the bite being received the spot where the fangs have entered begins to ache and swell. The feeling experienced is rather like the throbbing ache of a festering wound or abscess, and this increases as the swelling develops rapidly. If a small amount of venom has been injected the only symptoms that may develop are pain and swelling in the region of the bite. However, even in fairly mild cases of adder-bite the swelling is often extensive, and is an indication that the body's defensive mechanism is at work, rushing lymph, or body fluids, to the afflicted area, where they help to dilute the venom, thus reducing its toxicity, and also to inhibit its absorption into other parts of the body.

The swelling usually reaches its peak within forty-eight hours, during which time the throbbing pain gradually lessens, being replaced by a dull ache, and a feeling of heaviness in the bitten limb. Pain is frequently experienced in the glands in the armpit and the groin, but as a rule this is not severe and departs before the swelling has subsided. After the third day the swelling begins to subside slowly, although the bitten limb remains tender and painful when touched for several days afterwards. Loss of appetite is commonly experienced for two to three days following an adder-bite, being replaced by a considerable thirst, due to the dehydrating effect of the venom on the body. Discoloration of the bitten part is frequently noticeable from the beginning, but in most cases it is more evident as the swelling subsides, the skin in the region of the bite exhibiting successively the different colours of a resolving bruise. In some cases the only discoloration is a red, inflamed appearance of the swollen area. Complete recovery, without any lasting after effects, is usual within a fortnight.

In more severe cases a feeling of giddiness is often experienced within half an hour of being bitten, and this may be accompanied by vomiting. A feeling of drowsiness develops, with the victim eventually becoming comatose and partially helpless. A slight rise in temperature is usual, with resultant sweating and attacks of diarrhoea. Although perspiring, the victim often complains of feeling cold, and the swollen area around the bite frequently feels cold to the touch.

In very severe cases there is rapid collapse followed by temporary loss of consciousness. On regaining consciousness the victim sometimes complains of a feeling of constriction around the head or the chest. Breathing becomes laboured, and the swelling of both tongue and glottis makes speaking and swallowing difficult. Pain is frequently felt in the stomach, accompanied by severe vomiting and diarrhoea, and there may be loss of control over the limbs. Recovery in such cases may take some weeks, and the skin around the bite may eventually slough off in large pieces.

One of the effects of snake venom is lowered resistance to infection, but the risk of an adder-bite becoming septic is slight so long as no strong caustic solutions are employed in the first-aid treatment of the bite. It was at one time widely believed that the application of either ammonia or permanganate of potash to an adder-bite would neutralise the venom, but in fact the small amount of good these caustic solutions do is greatly outweighed by the destruction of tissue they cause. It was the use of these chemicals in former times that was often the cause of adder-bite wounds developing into running sores which were difficult to heal.

Over the past seventy-five years only nine deaths from adder-bite have been recorded in the British Isles. Six of these were children below the age of fifteen, and two of these were known to be in delicate health before they were bitten. Seven of the deaths occurred prior to 1945, and as far as is known none of them received treatment with antivenom serum. The remaining two fatalities occurred in 1957 and 1961; both were children, aged twelve and fourteen respectively, and both had been given antivenom serum.

In the nine cases of adder-bite which terminated fatally,

the time of death varied from between six to sixty hours after the bite had been received. There was no correlation between the age of the victims and the time of death following the bite. One three-year old child died within six hours of being bitten, whereas another child of four and a half years old lived for sixty hours. In two more cases a child of six years died thirty-six hours after receiving an adder-bite, while a child of eleven died in twelve hours.

On 17 April 1955 I experienced for myself the effects of adder-bite on an 'average, healthy adult'. I was on Arbrook Common in Surrey searching for adders, and it was, as I recall, a glorious sunny morning with the sun shining warmly from a cloudless sky. I had searched for over an hour without success when a rustling sound drew my attention to a tangle of dried grass that surrounded the base of a small bush. I turned just in time to see the latter half of an adder disappearing slowly among the bleached grasses. Moving quickly forward I laid the short stick I was carrying across the adder's back, grasped its tail with my other hand, and lifted it smartly from the ground. As it hung writhing head downwards, striking angrily in all directions, I could see at a glance that it was a young male adder, some 45 cm in length, with jet black markings on a greyish white background.

My attention was momentarily diverted from the adder as I endeavoured to tug my collecting bag from my pocket, when an extra loud hiss of annoyance from the adder preceded a sharp prick on my finger. Instantly I released my hold on the snake and it dropped to the ground, where I put my foot firmly but gently across its back to prevent its escape. With the adder thus secured I then turned my attention to the hand that had held the adder. Two bright red beads of blood on the third finger of my left hand indicated where the adder's fangs had entered. Hastily I reached for the razor blade which I always carried for such an emergency, and began making cuts through the fang punctures. The blood flowed freely, and I sucked vigorously at the cuts I had made in an attempt to draw out as much as I could of the venom.

Apart from the soreness resulting from the cuts I had

made I felt no pain in the bitten finger. I began to think that possibly I had succeeded in removing most of the venom, and that I might suffer no ill effects from the bite. Forgetting my bitten finger for the moment I set about the task of getting the adder off the ground again and into my collecting bag. Having accomplished this – with a great deal more caution than before, I might add – I became aware of another rustling sound coming from somewhere beneath the tangled grass near at hand. I at once began searching about with some enthusiasm, fully expecting to find the mate of the male adder I had just bagged. But after ten minutes or so of fruitless searching I became aware of a throbbing pain in the bitten finger, and on looking at it I saw that it was beginning to swell. Much as I would have liked to stay and find what I was sure was another adder somewhere amid the tangled grass I decided that it would, perhaps, be wiser to go in search of a doctor.

I had glanced at my watch shortly after being bitten and the time then was just past noon. It was now twenty minutes past as I set off towards the nearest road, which was about a third of a mile away. Even before I had reached the road the swelling had extended from my finger to my hand, which looked quite puffy and ached considerably. I paused to make further cuts to the underside of the now badly swollen finger in an attempt to reduce the swelling, and then continued on my way, wondering to myself as I walked along whether or not I would begin to feel faint or giddy. Fortunately for me these symptoms did not develop, and soon I reached the road where I managed to get a lift from a passing motorist to a doctor's house about a mile further along the road.

Approximately one hour after having received the adder-bite I was sitting in the doctor's surgery while he telephoned the nearest hospital for an ambulance. My hand was now about twice its normal size and the swelling had reached my wrist. With cotton wool and bandage the doctor's wife made a neat parcel of my hand, and then provided me with a cup of hot, sweet tea, for which I was most grateful, as my mouth felt unusually dry. By the time the ambulance arrived some fifteen minutes later my forearm had begun to swell and was throbbing and aching like my hand.

Shortly after my arrival at the hospital a tourniquet was applied to the upper arm just above the elbow, and a few minutes later a preliminary test shot of antivenom was injected into my right forearm; not the bitten limb. Fortunately for me there was just one phial of the required serum available at the hospital, and about half an hour following the test injection I received approximately 5 cc of this in each thigh by intramuscular injection.

I remained at the hospital until about 5 p.m., when I was discharged with instructions to rest. My arm was now swollen right up to the shoulder, and was at least twice its normal thickness. A slight pain had developed in the gland in the armpit of the bitten limb, whilst the throbbing ache initially experienced had lessened considerably. Apart from a somewhat inflamed appearance of the hand and arm there was no discoloration of the skin. Nor had I experienced any other symptoms since the bite was received, except for a sudden feeling of giddiness just before leaving the hospital which passed off quite quickly.

By the following morning the afflicted limb appeared to have swollen a little more. I had been asked to visit the hospital again that day, but as I did not feel in the least unwell I did not bother. I simply bathed the arm and hand with wych-hazel extract in an effort to alleviate the swelling, and also to cool the skin, which felt hot. The hand was a horribly bloated object, with the fingers as big as sausages. I kept the arm supported in the sling provided by the hospital because it felt almost too heavy for me to lift, and painful to move. It was also extremely tender and the slightest knock it received felt like a blow from a hammer.

The following observations on the symptoms experienced are from notes made at the time.

19 April. This morning my arm appeared no less swollen, but much of the pain has gone, and I am now able to move the arm without discomfort. The finger on which the bite was received does

not appear very swollen, and no pain is felt there. Fingers difficult to move due to swelling in the hand. Swelling worst near elbow joint. Slight pain that developed in the left groin is still there, and a slight swelling has developed at the front of the left ankle. I still feel very thirsty.

20 April. Arm less swollen this morning. Bathed it again with warm water and then with wych-hazel. Swelling subsided considerably during the course of the afternoon. I am now able to move the finger quite freely, and no pain is felt whatsoever. Nor is there pain in either the groin or the armpit. Swelling has now disappeared from ankle. Spent most of the day walking.

21 April. Swelling almost gone, except near elbow joint.

22 April. Very little swelling can now be seen. Slight pain felt in both hand and arm on exerting pressure.

26 April. Slight swelling remains around knuckle joint of finger bitten by adder. Mild irritation also felt there at times.

No further notes were made on the subject after 26 April. No after effects, or delayed symptoms, were experienced, and there is now no scar or mark on the finger that was bitten. The one thing that did surprise me was the complete lack of discoloration of the skin in the region of the bite, which is usually a significant feature of adder-bite. In many cases of adder-bite considerable discoloration of the skin occurs as the swelling subsides, but in my case this was entirely absent. In fact the symptoms I had experienced were comparatively mild, due partly, I imagine, to the small size of the adder, and possibly also because I had made a prompt incision of the bite. Possibly the fairly prompt injection of antivenom serum had helped in preventing the more unpleasant symptoms from developing, although I am inclined to believe that had the adder been a larger one, and therefore capable of injecting a larger dose of venom, such symptoms would have made themselves apparent before the serum was administered.

From the research point of view my own experience of adder-bite was somewhat disappointing, as it was not indicative of the adder's full potential, and wishing to discover for myself just what this might be, short of actual demise, I published letters in two prominent country magazines asking readers with first-hand experience of adder-bite to send me full details. I received a number of replies, both from people who had themselves been bitten by the adder, and also from those whose dogs had received bites. The most interesting of these are reproduced here, the cases of adder-bite in people being dealt with first.

The first letter is from a doctor living in Pontesbury, Shropshire, and concerns two cases of adder-bite, one mild and one very severe.

The patient, a healthy 14-stone man of about forty, was bitten on the hand. He reached the surgery about thirty minutes later in a really collapsed condition. He was given an injection of Piriton (an antihistamine) and without local treatment was sent to Salop Infirmary where he was treated with antivenom serum. He remained there for some days before making a complete recovery.

I have also heard of one other bite in a child of this area this summer, who was able to go back to school in the afternoon.

The next letter is from a doctor who was herself the victim. She writes:

In August 1942, when aged twelve years, I was bitten by an adder which I had picked up from the ground. The adder bit me in the web between the thumb and index finger of the left hand. Almost immediately afterwards my father made a superficial cut over the puncture wound with a penknife and sucked what he could from the wound. I felt perfectly well, and, after a swim and a picnic, walked the two miles home.

Swelling of the hand commenced soon after the bite and progressed up the arm. After two hours the dorsum of the hand was affected; subsequently the wrist, forearm and elbow. By the evening the arm up to the shoulder was tense, swollen and tender.

No antivenom serum was given. The only treatment I had was hot fomentations to the arm to attempt to reduce the swelling. The arm remained swollen for about two weeks, and when subsiding was coloured, blue, green, and yellow, like a resolving bruise. Some three weeks later the arm still felt tender and bruised. I have now no signs of abnormality or swelling in the arm.

I am now a doctor so in retrospect find this incident of great interest.

The following account was received from a woman who was bitten by an adder at the age of sixty-two.

Whilst out walking on neighbouring heathland on 21 March I felt a sharp pain on my ankle which I thought to have been caused by a bramble thorn. There was a small amount of blood on the ankle, which, after about two hours, began to swell and become stiff. I also found myself yawning constantly. At about 6 p.m. (bite was received at about 2.15 p.m.) intense pain was experienced in the calf muscle, and the instep of the bitten foot began to swell. This was followed by violent perspiration on the upper lip and brow, accompanied by two attacks of near fainting. I developed a great desire for water, and drank six to eight pints. For three days I ate no food, but slept quite well. I had no headache or rise in temperature.

On the morning following I was most violently sick after drinking a cup of tea. I had developed diarrhoea, but felt slightly better. My leg was much inflamed and very painful when put to the ground. At 10.30 a.m. I received an injection of penicillin from my doctor. On the third day the pain which I had experienced in my groin lessened, and the swelling began to subside. The instep of my foot and my leg were still swollen, and intense pain was felt on putting the foot to the ground.

After a fortnight the swelling had receded considerably, and after six weeks the bitten ankle was only slightly larger than normal, but slight pain was still experienced in that region.

The next account was supplied to me by a woman whose son, aged nine, was bitten by an adder on Arbrook Common in Surrey, which is a place I know well, having myself both caught and observed adders there over a number of years.

In 1948, my son was walking on Arbrook Common, near Esher, when he caught an adder with a forked stick. Picking it up rather clumsily he was bitten on the second joint of his index finger. I had told him if ever this should happen, there being many adders on both Arbrook Common and Oxshott Heath, he must suck the

32. A smooth snake attacks a slow-worm and endeavours to encircle it with its body. A fierce struggle ensues with the two reptiles rolling and writhing about on the ground. The thicker tail is that of the slow-worm.

33. With the slow-worm's neck held firmly in its jaws the smooth snake maintains constriction for almost twenty minutes, by which time the slow-worm's struggles have been quelled. The snake then moves its jaws sideways towards the slow-worm's head and begins the lengthy process of swallowing the legless lizard.

34. At this stage it almost looks as if the snake has taken on more than it can manage. It pauses now and then to rest its jaws, but afterwards continues steadfastly with the task of swallowing its enormous meal.

35. Infant smooth snake, six weeks old, feeding on baby slow-worm. While holding the baby slow-worm in its jaws the young smooth snake endeavours to constrict it by encircling it with its body.

36. Smooth snakes have been known to devour their own young. This photograph shows a baby smooth snake trying to swallow its own brother or sister.

37. Several days prior to casting its skin the grass snake becomes drab in appearance, and its eyes are obscured by a white opacity. The eyes clear within three to four days before the skin is shed.

38. By rubbing its head and jaws against rough objects the grass snake has succeeded in removing part of its old skin, which can be seen turned back around its neck.

39. Sloughing has almost been completed and the grass snake is on the point of drawing its tail out of the old skin, which can be seen in the foreground.

40. This young grass snake is shedding its skin for the first time, and within seventeen days of leaving the egg.

At the start of sloughing the skin becomes loosened around the head, and by rubbing its head against rough objects the snake succeeds in forcing the skin back towards the neck.

41. The old skin or slough has been partly removed and can be seen turned back inside out along the body.

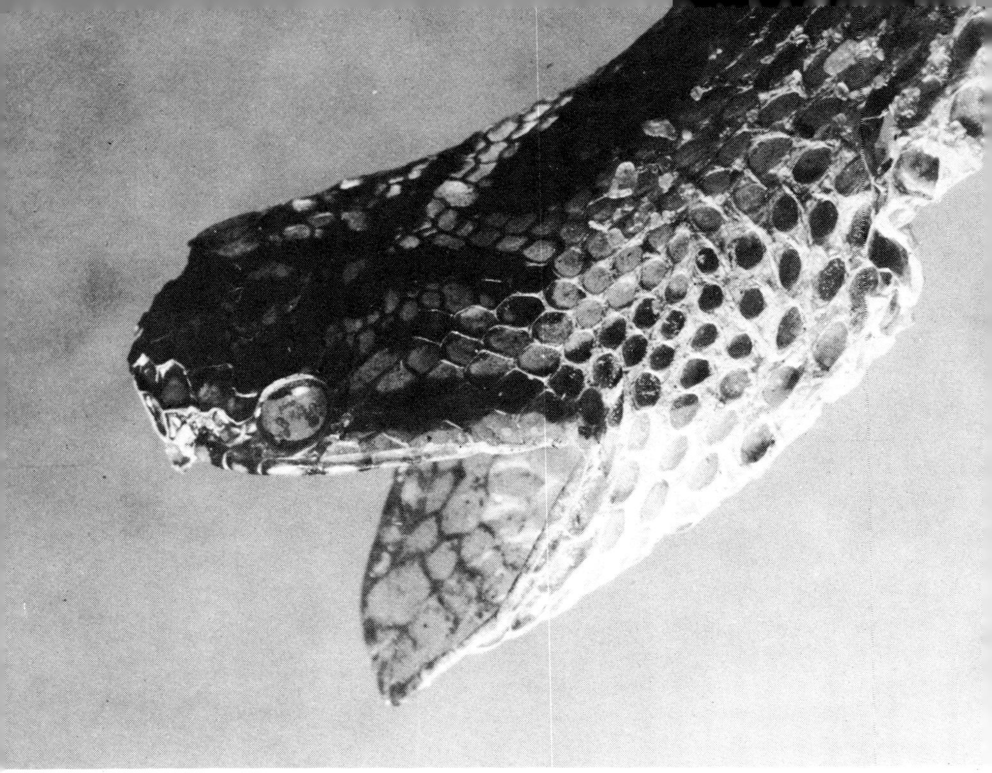

42. The head piece of an adder's cast skin, or slough, on which the distinctive inverted V-marking, and the start of the zig-zag band, are clearly visible. The transparent shields which cover the eyes can also be seen.

43. Grass snake yawns while basking.

44. Smooth snake basking. All three British snakes are in the habit of coiling themselves in symmetrical patterns, either elongated or circular in shape, when basking. This close-coiled attitude is only adopted when the snake feels completely secure and relaxed. It is an attitude often employed to conserve bodily warmth when the snake is basking in weak sunlight.

45. Male adder basking with flattened body. Both male and female will sometimes adopt this attitude when basking in weak sunlight. By doing so a larger area of the body is presented to the sun, and the snake is warmed more rapidly. Once the adder has become thoroughly warmed it reverts to its normal cylindrical shape.

46. A clearing among trees and bushes where bracken and long grass grow is a favourite haunt of the grass snake.

47. Typical habitat of both adder and smooth snake.

wound, spit, and then put on a tourniquet. In spite of being only nine years old at the time he remembered to do all these things and made his way home with all speed. I loosened the tourniquet and replaced it and then took him straight to Thames Ditton Hospital. Here great excitement prevailed as it was the first case of snake-bite they had had to deal with.

The doctor made cuts about half an inch long over the punctures and made the wounds bleed freely, he then re-applied the tourniquet and gave the boy hot black coffee. I made him walk about until he was tired and lay down on a bed.

Meanwhile the hospital contacted all hospitals within a radius of fifteen miles but none of them had a stock of anti-snake-bite serum. They also contacted the Reptile House at the London Zoo for advice.

My husband was telephoned in London and instructed to go to Messrs Allen & Hanbury's, to obtain snake-bite antidote and bring it down with all speed by car.

Very strangely the dog of some friends of ours living directly opposite had been bitten on the nose by an adder a few days previously. The dog's nose had swollen up to a fantastic size, but they had obtained some anti-snake-bite serum, I do not know from where, and the dog had been injected with half the bottle. The son of these friends offered us the remaining half bottle which had been kept in a refrigerator. I sent a taxi to fetch it and this was injected into my son pending the arrival of the supply from London. Meanwhile the child's arm had swollen to an enormous size from the fingers, which were like a bunch of sausages, to the elbow, and the whole of the arm was almost navy blue in colour. I am inclined to think that this was owing to the tightness of the tourniquet, which gave him considerable pain. The gland in his armpit was also painful. When the further supply of serum arrived he was given a large dose, and the tourniquet removed. After a night in hospital he came home and was a little dopy for a day, but when the swelling had subsided after about four days he seemed none the worse, and only the scars remain to this day to tell of his adventure, which was made much of in the local press.

Among domestic animals the dog probably falls victim to the adder more frequently than any other, chiefly because it is less confined than most. Furthermore its natural curiosity, combined with its habit of questing about in search of some animal to hunt, increases the chances of its getting bitten by the adder.

Gun dogs are particularly vulnerable to adder-bite, as their work entails seeking out game in dense cover, and often in the sort of places where adders are most likely to be found. Letters I have received from people who have had personal experience of adder-bite in dogs confirm this, as in those places where the adder is common gun dogs frequently get bitten, some of them receiving bites on more than one occasion.

One gamekeeper living in Ross-shire wrote telling me that he had personally dealt with about twenty cases of adder-bite in gun dogs, and that in the majority of these the dog had been bitten on the legs during the grouse-shooting season. Apparently none of these cases proved fatal, and most of the dogs recovered completely within about five days.

Other letters received told of dogs, not all of them gun dogs, receiving bites on the nose, lower jaw, and, in the case of a dachshund, on the stomach. One letter came from the owner of a cocker spaniel, which, it appears, made a sport of hunting out and baiting adders. In consequence it had received bites on three separate occasions, none of which was very serious. The dog eventually gave up this pastime, no doubt deterred by the painful outcome of jousting with poisonous snakes.

About seven months before I was bitten by an adder on Arbrook Common, my crossbred terrier received a bite from an adder on the very same common. The adder that bit him was a young female 40 cm in length. Its fangs punctured the dog's muzzle just above the nose while he was poking about among some long grass. He let out a yelp of pain and jumped back so quickly that he fell over backwards. I immediately made two small incisions with a razor blade through the fang punctures, and despite the dog's growls of protest, sucked hard at the wound. I then set off with the dog at heel in the direction of the nearest road, which was about a quarter of a mile away.

Within ten minutes or so of receiving the adder-bite my

dog was in a very dazed state, and even walking very slowly appeared to be great effort to him. He kept wanting to stop and rest, and eventually I had to pick him up and carry him for most of the way to the road. He weighed only forty pounds, but in a temperature of 26·5° C in the shade (it was early September and unusually hot) it was warm work walking over rough ground with him in my arms. Eventually we reached the road where a passing motorist very kindly took both dog and me to the nearest vet. There, within an hour of being bitten, the dog received an injection of anti-venom serum in the neck. His muzzle was then only slightly swollen, as was also his neck. Nevertheless, the swollen parts must have been very tender, for when touched on the neck he responded with a growl, and it was necessary to bandage his jaws together while the serum was administered.

Immediately after the injection of serum I took the dog home, and on arrival there he went straight to a dark corner where he remained, scarcely moving at all, for the rest of the day. A water bowl was placed near him, and it seemed as much as he could do to lift his head and drink from it, which he did frequently and copiously. To my great surprise he ate his evening meal as usual, making his own way into the kitchen to do so, and then retired again into his dark corner to rest.

By the following morning both his muzzle and neck were badly swollen, and his head looked about twice its normal proportions. I found him still lying in the same corner, and he made no effort to get up when I approached him. He spent all that day resting, making no attempt whatsoever to move, but drinking frequently from his water bowl. I bathed his muzzle and head several times during the day with wych-hazel to ease the swelling, and by the evening he seemed a little better. He ate his usual meal and afterwards he appeared less lethargic. Later in the evening he managed to summon up enough energy to accompany me on a short walk. By the third day the swelling had subsided considerably, and on the fourth day following the adder-bite he was practically his old self once more.

When I first arrived at the vet's house I had expressed my concern for the dog because of his age (eight years), and the fact that he was a little overweight. The vet assured me that he would most probably recover, for she had recently treated a smaller dog for adder-bite that was 'swollen like a balloon' before the antivenom serum was injected into it.

To my great relief the vet's prediction proved correct, and my dog lived for another six years without suffering any after effects as a result of the adder-bite.

From amongst numerous cases of adder-bite in dogs reported to me by their owners I have selected three which I consider to be the most interesting. The first comes from a correspondent living in Northumberland.

> The worst case of adder-bite in a dog that I have experienced occurred some eight years ago, when my favourite cocker spaniel got bitten on the nostril. I gather he must have gone up to the adder and sniffed at it. His head became swollen at once and he had to be carried a mile home. By next morning it was no exaggeration to say that his head was about two and a half times its normal size, and he couldn't raise it at all. On the vet's advice, we bathed the wound every half hour with permanganate of potash, and also gave him weak doses of whisky at intervals. It was about three weeks before he recovered and could walk about normally.
>
> This spaniel is still alive at the age of nearly fifteen, but some years ago he went stone deaf, and I wondered if the bite was responsible for this.
>
> My present labrador has been bitten twice. The first time was not serious because I don't think the adder caught him properly. But four years ago, when I was at one of my farms, he was badly bitten on the hind leg. We brought him straight home in the Land Rover, but it was most difficult getting him out when we arrived, because he seemed completely paralysed in the hind quarters and he was very ill for a week.

The next account of adder-bite concerns a sealyham terrier that was bitten on the nose. The owner living in Shropshire writes:

Many years ago my sealyham terrier was bitten on the nose. We found two small fang marks just above the black tip of his nose. My mother instantly sucked the place, and spat out as much as possible, but she said that the roof of her mouth went dead, as when a dentist gives one an injection before extracting a tooth, and remained dead for some hours. Despite this the dog swelled considerably all over his face and down his neck and throat. Having no permanganate of potash we applied all alkalis which came handy: soap, ammonia, and whisky. In about eight hours the swelling began to go down, and in roughly a couple of days the dog was completely normal.

The numbness of the mouth referred to in the foregoing letter was doubtless caused by adder venom drawn from the bite entering abrasions within the mouth. Venom can also enter the bloodstream through decayed teeth, which is one of the risks attendant upon this type of first-aid treatment.

The third and last account of adder-bite in a dog is of particular interest, since the bite was received not on the muzzle or the legs, as most usually happens, but on the stomach. With a short-legged dog such as the dachshund it would not be difficult for the adder to deliver a bite to that part of the body, although it seems to suggest that the dog was unaware of the adder's presence, and possibly stepped on it. If the dog had either seen or scented the adder first it would more probably have been bitten on the nose.

On 25 September my dachshund bitch (aged five years) was bitten on the stomach, just behind the ribs, by an adder. I carried her back to the car, about a mile or so, and by the time we reached it she had swelled considerably around the bite, and seemed almost comatose. Being Sunday evening I was unable to get the vet and the chemists were closed, so there was nothing I could do. She was very sick during the night and moaning as if in much pain. First thing next morning I took her to the vet who said it was no use giving her the usual injection for snake-bite at that advanced stage, but instead gave her a strong injection of penicillin. She was able to walk that day. She felt stone cold to the hand, and was enormously swollen even to the paws. The day after that she had a high temperature. She would drink a little but not eat, and seemed to have difficulty in swallowing. On the fifth day she seemed better and less comatose. She received penicillin for six days. She got on well, but in ten days time an abscess appeared where the bite was, which caused a great deal of trouble and spread practically all over the stomach, terrible masses of matter coming from it. Eventually the entire skin of the stomach peeled off in thick pieces. It was many weeks before it healed, otherwise her health was good. She became one hundred per cent all right after a long time.

CHAPTER ELEVEN

Treatment of adder-bite

Pros and cons of antivenom – production of antivenom serum – alternative measures – advice to adder-bite victims

FROM 1936 until about 1962 antivenom serum was the accepted form of treatment for adder-bite in this country. However, shortly after the second of two fatalities from adder-bite, which occurred in 1957 and 1961 respectively, following treatment with antivenom, the British Red Cross Society approached the Royal Society of Tropical Medicine for advice on the treatment of snake-bite. In consequence an inquiry was made into the treatment of snake-bite, which resulted in a statement being published in both the transactions of the Royal Society of Tropical Medicine and the Monthly Bulletin of the Ministry of Health that the bite of *Vipera berus* was less dangerous than the use of antivenom. It was also stated that the only antivenom available was produced from the venom of *Cerastes cornutus*, a South African viper. This was later amended to read, *Aspis cerastes*, a North African viper.

The only antivenom available at that time for the treatment of adder-bite in this country was ER Antiviperin, which is produced by the Pasteur Institute in France. It had been imported into this country for more than a quarter of a century when, in April 1964, the Ministry of Health decided that not only was its effectiveness against the bite of *Vipera berus* unproved, but that it also did not comply with the requirements of the Therapeutic Substances Act 1956, and its commercial import was stopped. However, ER Antiviperin can still be imported by hospitals where it is felt there is a local need.

My own inquiries into this matter have revealed that medical opinion is divided on the question of whether or not antivenom serum is an effective antidote to snake-bite. I also learned from the former importers of ER Antiviperin that this serum is produced from the venoms of *Vipera berus* and *Vipera aspis*, and that the antivenom produced from the African viper *Cerastes cerastes* (formerly known as *Aspis cerastes*) is an entirely separate one that has not been used in this country for treating the bite of *Vipera berus*.

Ironically enough the Royal Society of Tropical Medicine have published in their transactions (vol. 60, pp. 797–810, 1966) a list of antivenoms in which it is shown quite clearly that the antivenom produced by the Pasteur Institute as an antidote to the bite of *Vipera berus* is manufactured only from the venoms of European vipers.

It might be thought by some that adder-bite in this country is of such rare occurrence that the question of its treatment is of little significance. However, it has been stated that in 1961 alone there were four hundred requests made for antivenom serum.

The case against ER Antiviperin – excluding the erroneous statement concerning its manufacture – is that it is alleged to be a relatviely unpurified horse serum, which is so high in non-specific protein that it could cause protein, or anaphylactic, shock in those allergic to serums produced from horses, and could prove fatal. However, a point which I feel has not been sufficiently stressed is that the risk of anaphylactic shock can be avoided by making a preliminary test injection of a small quantity of antivenom to ascertain whether or not the adder-bite victim is allergic to the serum. It might be argued that in some cases the urgent need for treatment would preclude a preliminary test being made. As, however, the adder is not a highly venomous snake, such urgency is unlikely to arise, except perhaps where the victim is a very young child. Serum sickness can, of course, develop several days after an injection of antivenom, but this, although unpleasant, is not normally dangerous and eventually passes off.

Antivenom serum is produced by injecting an animal, usually a horse, with increasing doses of snake venom over a period varying from between two to seven months. The process is so gradual that it has little or no harmful effect on the horse, which develops antibodies to counteract the venom and eventually becomes immune to it. When this stage has been reached lymph, or body fluid, is taken from the horse and processed to make antivenom. Processing consists mainly of removing, or attempting to remove, from the lymph or serum certain elements which have not produced antibodies in the horse.

Snake venom is composed largely of various proteins and enzymes, some of which are not good antigens, meaning that they do not stimulate the production of antibodies to themselves when injected into an animal, such as the horse, for the purpose of producing antivenom. It is these poor antigens which form the harmful components of antivenom serum, and which are responsible for producing adverse reactions in some snake-bite victims.

There is now available in this country a new antivenom known as 'Europe', manufactured by Behringwerke AG in Germany, which can be used for treating the bite of *Vipera berus*, being also an antidote to the venoms of *Vipera ammodytes*, *Vipera aspis*, *Vipera lebetina*, and *Vipera xanthina*. As it is produced from the venoms of more than one species of European viper it is no more specific than ER Antiviperin. However, the manufacturers of 'Europe' claim that improved processing has reduced the risk of sensitisation to horse serum. Nevertheless, they are careful to point out that 'before giving the serum it is essential to exclude anaphylaxis or allergy by careful questioning or testing or both'.

For adder-bite victims allergic to antivenom there are alternative methods of treatment, and one which is favoured by some members of the medical profession in preference to antivenom involves the use of antihistamine, antibiotic, and anti-tetanus injections. But even with this form of treatment adverse reaction is a probability, as there are many people who are allergic to these drugs.

Medical opinion also appears to be divided on the effectiveness of certain other forms of treatment for snake-bite which were formerly thought to be of value. These include incision of the bite, and the application of a tourniquet to localise the venom. Some advise the use of a tourniquet, and incision of the bite to relieve pressure which might result in gangrene, while others are of the opinion that neither treatment has much value, and may increase the risk of infection and gangrene in the region of the bite.

A point worth noting is that in many severe cases of snake-bite where the victim has not responded to treatment with either antivenom or drugs, a blood transfusion has usually brought about a sudden and marked improvement in the victim's condition.

As to the question of first-aid measures, the soundness of some of the advice offered to victims of snake-bite in this country by certain sections of the medical profession is open to question. It is frequently recommended that the snake-bite victim should, if possible, kill the snake and take it with him to the hospital for identification. Personally I would not advise anyone to attempt to kill the snake that has bitten him, as in doing so he runs the risk of receiving further bites. Snakes are notoriously difficult to kill, and snakes that have been thought to be dead have been known to deliver a poisonous bite, as even those experienced in handling snakes have sometimes discovered to their cost. The decapitated head of a rattlesnake has been known to deliver a poisonous bite forty minutes after having been severed from the body.

Retaining the dead snake for identification purposes is pointless, since the adder is the only poisonous snake indigenous to Britain, and anyone bitten by it would soon be made aware of the fact, as within ten minutes or so of the bite being received the bitten part would begin to swell rapidly, accompanied by a throbbing pain.

The only other British snake likely to bite if picked up or

molested is the smooth snake, which is very rare and therefore unlikely to be met with. Being non-venomous, and possessing only very small teeth, its bite produces a few minute punctures in the skin and nothing more. The remaining British species, the grass snake, is also harmless and, in my experience, very rarely bites. The chances of meeting up with a foreign species of poisonous snake that may have escaped from captivity are extremely slight, as such snakes are not normally kept as pets, and specimens imported into this country are only made available to accredited herpetologists.

The best advice that can be offered to anyone unfortunate enough to be bitten by an adder is that they make their way at once to the nearest doctor or hospital, without wasting time trying to kill the snake. In any case there is little likelihood of the adder-bite victim capturing the offending snake either alive or dead, as having done its worst it would quickly make itself scarce.

As exertion of any kind speeds circulation of the venom throughout the body, the adder-bite victim should, where possible, be taken by some form of transport to obtain medical aid, rather than be allowed to walk. A child should be carried if transport is not available. Treatment of the bite is best left to those medically qualified, as first-aid treatment by untrained persons may do more harm than good. Hot drinks of strong tea or coffee may help to counteract shock, which is a common symptom following snake-bite. But on no account should alcohol be given as this can have an adverse effect. It is important that the victim be reassured that death from adder-bite is extremely rare in this country, as fear of the consequences can have a deleterious effect.

There is, of course, always the possibility that a person stung by some kind of insect might believe that he has been bitten by an unseen snake. In some cases the symptoms may at first appear similar, although the swelling produced would be much less extensive. In such cases the best advice that can be given is that the victim, if in doubt, should seek medical aid.

CHAPTER TWELVE

Observing snakes

Snakes' stealthy behaviour – time and place – use of eyes and ears

SNAKES are not easily observed in their natural environment, for unlike many wild animals which can be studied at their nest, burrow, den, or lair, snakes are creatures of no fixed abode. The snake's hiding-place – one can hardly call it a den – is usually near to where it basks on sunny days, and if disturbed too frequently by intruders while basking the snake will readily remove itself to another spot. A favourite hiding-place is among the roots of a small bush, although woodpiles, heaps of dead bracken, and the disused burrows of other animals are also used.

Many wild animals can be more easily observed during the breeding season, but this is not the case with British snakes, which do not make nests or dig burrows in which to produce their young. The grass snake forms a cavity in loose soil or in a compost heap wherein to lay its eggs, but it does not remain with the eggs after they are laid. Neither adder nor smooth snake provide any shelter whatsoever for their offspring, which are usually born under cover of surrounding vegetation, and not in any particular place. Consequently it is extremely difficult to study the breeding behaviour of snakes in their natural habitat. This also applies to the snake's feeding behaviour, as it is unusual to come upon a snake feeding. I have watched on occasion a snake hunting for food, but one rarely has the opportunity of seeing a snake make a capture, and prey, once caught, is invariably taken under cover to be swallowed. Nevertheless, observations on snakes living wild have been made, but more by chance than design. One could stand and watch a snake basking in the sun for hours without seeing anything of interest happen.

In places where snakes are fairly numerous one may sometimes be fortunate enough to observe the mating behaviour, as during the mating season snakes are slightly less timid, and due to the lack of dense herbage at that time they are more easily seen, but such opportunities are very infrequent.

Before one can hope to study snakes outside captivity one needs to know where to find them. Many places which seem ideally suited to a particular species may often prove to be quite devoid of them. By continually searching for snakes one learns to recognise the sort of places they choose to inhabit. The more unfrequented and lonely the locality the more likelihood there is of it being the habitat of snakes, for their greatest dislike is having their solitude disturbed, particularly by man.

All three British species are in the habit of basking close to cover of some kind into which they can dive for safety when danger threatens. Sometimes a snake will be found lying on bare ground close to a bush or large tuft of grass, but snakes most usually bask in small clearings among dense grass, thick heather, or close to growing bracken. Any surface that holds the sun's heat is a favourite basking place, and for this reason small logs and heaps of dry bracken are frequently chosen. Low mounds, such as molehills and anthills are commonly used as basking places, as are also large tufts of grass, provided there is dense surrounding cover into which the snake can retreat if necessary. It is unusual to find a snake out in the open away from cover, for example, on close cropped grass where there are few bushes, or on heathland where the ground is only sparsely covered with heather.

If one is to be successful in seeing snakes the weather must be warm and preferably sunny. On days when the sky is overcast and the air cool there is little chance of finding snakes out and about. Snakes do not like strong winds, and even in sunny weather they will not be much in evidence if it is very windy. A light southerly wind on a sunny day will not prevent snakes from basking, whereas northerly or easterly winds make them stay under cover, except in very sheltered places.

The ideal weather for a snake hunt is when the sky is cloudless and the air calm. In early spring such conditions will draw many snakes from their hibernacula, and whilst the warmth from the sun is sufficient to induce them to bask, the temperature does not rise high enough to make them fully alert, so that they are more easily taken by surprise. Later in the spring and throughout the summer the higher temperatures that usually prevail make snakes quicker in their reactions, and able to sense more readily the approach of an intruder.

From the beginning of May until late August, a clear sky and a still air invariably result in a rapid rise in temperature after sunrise, and by midday the sun is often too hot for snakes to remain out basking. For although snakes love the sun they do not like intense heat. Of the three British species the smooth snake is the best suited to high temperatures, while the grass snake can sometimes be found basking when the adder has already sought the shade. During the summer months it is, therefore, best to search for snakes between 7 a.m. and 11 a.m. If one begins searching too soon after sunrise one may, by moving about, prevent snakes from emerging from their hiding-places. After 11 a.m. it is usually a waste of time and energy to continue searching, for by that time the sun's heat will have driven most snakes into the shade. During spells of very hot weather snakes spend only a little time basking, and usually within an hour or two of sunrise. If not basking at this time they will probably be out hunting for food, before the sun becomes too hot.

Although snakes show a preference for sunny weather, they may also be found lying out on close, thundery days when the air is very still and the sky veiled by heat haze, through which the sun shines only faintly. I have found snakes lying coiled in a basking position in such weather, even though there was little direct heat from the sun. A sunny day in early autumn is also a good time to see snakes, for then, as in the spring, the temperature does not rise very high, and snakes can lie basking throughout the day without getting too warm. Indeed, most snakes seem to make the most of this last opportunity to sunbathe before frost and cold winds drive them into hibernation.

Equally as important as warm, sunny weather to the successful observation of snakes in their natural habitats is the ability to be able to move quietly. For although snakes do not possess ears, and cannot hear in the way we do, they are extremely sensitive to ground vibrations and can in this way detect the approach of an intruder many yards distant. It is therefore essential to tread softly. To be able to do this successfully one must move very slowly, as it is well-nigh impossible to move quickly over the uneven terrain where snakes are to be found without tripping or inadvertently putting a foot down heavily from time to time. Furthermore, by walking slowly one has more time to look carefully all around, and by so doing there is less chance of missing the snake that may be lying quite still, waiting for you to pass by before it sneaks stealthily under cover. But no matter how carefully you walk it is almost certain that you will see more snakes' tails vanishing rapidly beneath a bush or into a clump of grass than you will see snakes basking unaware of your presence.

In addition to walking slowly and carefully one should stop frequently and scan the ground all around. By standing still and looking there is more likelihood of your seeing a snake before it catches sight of you. For although a snake's vision is not good it is quick to spot movement. From the snake's low viewpoint a man must tower up against the sky in much the same way as a huge prehistoric monster would appear to us, and any movement of such a large creature against the clear background formed by the sky would be quickly spotted. Even with the comparatively keen eyesight that we ourselves possess it is not easy to locate an animal that remains absolutely still, whilst for the snake with its poor eyesight it must be much more difficult. Thus by keeping immobile it is often possible to observe without being observed oneself. For the same reason a snake will often remain unmoving although it is aware of the presence of an intruder, for it knows that by doing so it may escape detection. It is in fact possible to stand within a few feet of a basking snake and not see it until it moves, although this happens less frequently as one becomes more adept at snake hunting.

Often one becomes suddenly aware of a snake's presence at the very moment that it becomes aware of yours. If your approach has been slow and cautious it is likely that the snake will not make a sudden dash for cover, but after a short pause glide slowly and stealthily away into the herbage. As the snake is not greatly alarmed the best thing to do is to remain absolutely still and wait. After about twenty minutes have passed the snake may re-appear if it feels safe. But often the snake does not appear again, no matter how long you may wait. This may be due to the snake being able to detect your scent, in which case it will not show itself until you have departed. By standing downwind of where the snake is hiding there is less chance of it winding you, although even on a calm day the snake may be able to scent your presence if you are fairly close, for snake's possess a very keen sense of smell. However, another reason why the snake may not show itself is that it may not have remained in hiding near to where it was last seen, but made off very slowly and quietly to another spot. Snakes are very adept at creeping stealthily away in this manner after they have been disturbed, which is another reason why they are not easily observed in their natural habitat. Quite often when I have wished to capture a specimen that has dived into cover at my approach I have searched around for some time without finding it. The snake is most probably coiled up among the roots of the grass or heather near to where I am standing, but trying to find it is like searching for the proverbial needle in a haystack. However, by keeping still for a while and listening intently I have sometimes been able to hear the slow, almost imperceptible rustling sound made by the snake as it makes off with the utmost stealth under cover of the herbage. Sometimes a snake makes its escape both unseen and unheard by making use of mouse runs which often travel for long distances just beneath the surface of the soil. Snakes frequently travel along these narrow tunnels, particularly when being pursued, which explains how they are sometimes able to vanish so mysteriously.

When searching for snakes one should not only look but also listen, for often the first indication of a snake's presence nearby is the rustling sound it makes as it glides away. If alarmed and moving quickly the sound is plainly heard and is quite distinctive, being a continuous rustling as opposed to the intermittent rustling made by lizards, and most other small creatures, which move more hesitantly through the undergrowth. I have located as many snakes by sound as I have by sight. I can recall one occasion in particular when I located three pairs of mating grass snakes by the sound they made moving about amongst long grass. Each pair was found in a different spot, but within a matter of yards of one another.

Another way of detecting the presence of snakes is that of watching for movement of the herbage. Tall grass will sway gently and continuously if a snake is moving slowly through it, although such movement can only be detected on a calm day. Most other small creatures move more jerkily and as a result the swaying of the grass starts and stops abruptly.

Whilst searching for adders one May morning on a common in Surrey, I stood quite still for some time, listening for any sound that might betray the presence of a snake. Suddenly I noticed the tops of the tall grass some yards away swaying quite violently. As I continued to watch I realised that whatever it was that was causing the disturbance in the grass was moving in my direction, and at quite a fast speed. Presently two adders came dashing side by side towards me, and in my eagerness to get a good look at them I moved. Immediately the two adders separated, one vanishing into the grass on my left, whilst the other stopped half hidden in the short heather within inches of my feet. I had caught only a fleeting glimpse of the one that went off to my left, and so I was unable to identify its sex, but the one at my feet was undoubtedly a male. Whether or not the adder near me was aware of my presence at that moment I was not sure, but it made no attempt to move. Possibly it realised how close I was to it and thought that it might escape detection by keeping still. Slowly I bent down and caught hold of it by the tail, the only part of it that I could see clearly. The adder immediately hitched itself on to the heather and tried to pull itself free, but as I was anxious to get a better look at it I did not release my hold. The next instant the adder's head whipped round, and I let go its tail only just in time as its poison fangs skimmed passed the tip of my thumb. A moment later the adder vanished beneath the heather, and I was unable to find it. I regretted afterwards having attempted its capture, for had I remained quite still the two adders might have continued with their courtship or rivalry, whichever it was I had disturbed. It could have been either, as two males will often race side by side in a display of rivalry, whilst the same thing occurs between male and female during courtship. To be able to distinguish one form of display from the other one needs to be able to identify the sex of both snakes, and this I had been unable to do since one of the adders had vanished too quickly.

On another occasion, when I was accompanied by my nephew on a snake hunting expedition on Elstead Common in Surrey one morning in early May, I heard a rustling among the dry grass and heather near to the path we were following. Telling my nephew to remain absolutely still we waited and watched, and within a few minutes a male adder came into view. Moving unhurriedly through the tangle of grass and heather the adder eventually came to within a foot or so of where we stood. It then moved a little further away, as though it were about to cross the path on which we were standing, but changed its mind and moved again in our direction, passing once more very close to our feet. It paused for a while, its tongue waving slowly, then continued on its way, gliding effortlessly through and over the dense herbage. For some ten minutes or so we watched the male adder gliding back and forth over the same piece of ground, while gradually moving further from us.

I decided to follow the adder, but found it difficult to do so quietly among the tangle of fallen sapling trees and small bushes still bare of leaves. Although I was moving somewhat clumsily, and on more than one occasion tripped and jarred the ground with my feet, the adder showed no sign of alarm, although it was seldom more than a few feet away. Even more extraordinary was the fact that the adder reacted to my clumsy movements not by dashing off or diving into cover, as an adder would normally do, but actually turned and came towards me, as though to investigate the cause of the disturbance. On one occasion it came within a foot of where I stood, and simply glided by as though completely unaware of my presence. As I was endeavouring the whole time to obtain a photogrpah of the adder, I was continually moving about in an attempt to get an unobstructed view of it. Yet that adder continued to behave as though I were not there. It appeared to be searching for something, most probably a female, as it was then the mating season. I got the impression that the disturbance I was creating made the adder think that there was another adder near by, either a female or a rival male, which was why it came towards me instead of moving away, as would have been its normal behaviour at any other time of the year.

With practice one learns to move with almost the stealth of a cat when hunting for snakes, and thereby get close to them without being detected, although this is not always easy when searching for snakes amid a tangle of vegetation and fallen timber such as I have just described. Nevertheless I have on numerous occasions been able to get within two or three feet of a snake lying coiled in the sun, and stand watching it for some time before it has become aware of my presence.

One way to observe snakes during the mating season is to take up a position near to where snakes have been seen before and simply watch and wait. If more than one mating pair have been seen at a particular spot in a previous year it is just possible that there is a communal hibernaculum somewhere nearby, in which case there is every chance of seeing snakes there each spring. Alternatively, it may well be that snakes gather in that particular spot each year for the purpose of mating. Having discovered such a place it may be possible to observe them there in successive years.

As I have stated more than once already in this chapter, snakes are not easily studied in their natural environment, and detailed observations of certain aspects of their behaviour are not possible in the wild. I have, therefore, frequently found it necessary to capture specimens for close study in captivity. I should like to stress here that such specimens are rarely kept captive for more than a few weeks before being returned to the place of their capture. In the case of the adder, which often refuses to feed in captivity, it is often necessary to release specimens within a fortnight of capture. A long stay in captivity could make a snake less cautious when returned to its natural habitat, and this would be detrimental to its own safety. For the same reason no attempt is made to tame the snakes while they are captive, and they are handled as little as possible.

I have found by keeping snakes under conditions made to resemble as closely as possible their natural habitat, and by not permitting them to see their captor while they are under observation, that they tend to behave in much the same way as they would when living free.

Having captured a snake it is necessary to have something in which to carry it home, and I have found that the ideal container, which can be easily carried in a pocket when out on a snake hunting expedition, is a deep cloth bag. This is made from a piece of sheeting, or any other available material through which air can penetrate. The bag needs to be at least 60 cm deep and about 25 cm wide, with tapes for fastening stitched about 15 cm from the top. Once the snake is inside the bag the tapes are tied tightly around the neck of the bag to prevent its escape. For adders I use a bag with a wire ring sewn into the open end so that I do not have to fumble to get the reptile into it. Double fastening tapes are stitched a few centimetres below the wire ring.

Without the wire ring it is difficult to hold the bag open

with one hand so that the adder can be dropped into it with the other hand. The idea of using a ring of wire to hold the collecting bag open occurred to me some years ago when I was in the habit of capturing adders with my bare hands. I had found that the hand holding the collecting bag was the one most likely to get bitten as the adder, held suspended by the tail in my other hand, was lowered into it. The adder would usually be striking wildly in all directions, and my efforts to get it into a collecting bag that would not stay open often resulted in the snake's poison fangs narrowly missing the hand holding the bag. With the wire ring in position the adder could be speedily slipped inside with much less risk of a poisonous bite resulting. But no matter how experienced one may be, or how careful, in handling poisonous snakes there is always the possibility of a bite being received if gloves are not worn. I was reminded of this only this year when, as I was about to drop into my collecting bag a large male adder I had just captured, it struck suddenly and plunged its poison fangs into the thumb of the hand that was holding the bag. Fortunately my hand was protected by a stout leather gauntlet, which saved me from a second experience of adder-bite, which might have proved more unpleasant than my first. A point of some interest was that the adder had at no time tried to strike at the bag I was holding, and all its previous strikes had been aimed at my hands.

A short handle attached to the wire ring in the top of the collecting bag would undoubtedly be an improvement, as then an adder would have little chance of biting its captor's hand.

Before I began using the cloth bags described I used a deep metal can with a wire handle for containing the snakes I caught. But the very first adder I captured was carried home in a flat, round tin that had originally held sweets and which was no more than 8 cm deep and about 25 cm across. The capture of that first adder took place some twenty-four years ago, but I can still recall quite vividly the excitement of that occasion.

Accompanied by my terrier I had spent most of a warm

and sunny spring day exploring the wide expanse of what was then known as Thursley Moors, but now called Thursley Common. By late afternoon I had reached Pudmore Pond on Elstead Common, which adjoins Thursley Common, the two commons forming a large, continuous stretch of heathland some three miles long and about a mile wide. Most of this expanse of wild country is covered with heather, much of it growing on boggy ground with the wetter places marked by cotton grass. Clumps of trees, mostly conifers but with some birch, are scattered about the heath, which is mainly flat and open, rising in places to slightly higher ground, parts of which are covered with bracken interspersed with heather.

The purpose of my exploration was to find an adder, but although I had been searching all day, mostly in the wrong places, I had met with no success. It must have been around 4 p.m. as I picked my way over the marshy ground surrounding Pudmore Pond, and the sun was lowering in the cloudless sky. Having been unable to find an adder I thought I might be lucky enough to find natterjack toads in the vicinity of the pond, this particular part of the country being one of the few places which they are known to inhabit.

As I drew closer to the edge of the pond the ground became increasingly waterlogged, and I found that I had to jump from one large tussock of grass to another to avoid getting my feet wet. Strangely enough those large grass tussocks no longer seem to grow on the fringe of Pudmore Pond, but they most certainly existed there then, as I recall only too well. For suddenly, and quite unexpectedly, I caught sight of a female adder of average adult size lying compactly coiled on the top of a tussock of grass only a few metres from where I stood perched on another grass tussock.

I was so surprised that for a moment I did not quite know what to do. I wanted to capture the adder but I had no gloves with me, nor a stick with which to pin the reptile down while I grasped its tail. I could not see a suitable stick within reach, and as there was no time to spare to look for one, I had to make do with a dead twig I found lying nearby. I then began to stalk the adder as stealthily as I knew how, and eventually

I got within about a metre of it without causing it alarm. Shaking with excitement I edged closer and pressed my dead twig across the adder's neck, and immediately the twig snapped. The adder at once began to move off, but I succeeded in pinning it down once more with my now somewhat shortened twig, and grasping the reptile by the tip of the tail I lifted it quickly from the grass tussock so that it hung head downwards.

I had read of this method of capturing an adder in W. H. Hudson's, *Book of a Naturalist*, but this was the first time I had put it into practice. According to that famous naturalist the adder, when picked up in this fashion, could not reach up to bite the hand that held it, and I recall hoping at the time that his advice was trustworthy. Fortunately it was, but although the adder was unable to bite me whilst being held by the tail, my biggest problem was getting it into the ridiculously small tin I had brought with me for the purpose of transporting an adder home should I chance to capture one. Prior to capturing the adder I had removed the lid of the tin to facilitate getting the adder into it as quickly as possible, but the problem now was to get the lid on again without getting bitten once the adder was inside. I realised that I would have to be very quick in replacing the lid of the tin if the adder was to be prevented from immediately slithering out again. Alas, I performed the operation with such speed that the poor adder's tail was trapped by the lid, and the unfortunate creature promptly responded with an angry hiss. Gingerly lifting the lid I succeeded in allowing the adder to withdraw its bruised appendage without escaping.

With the adder now firmly secured within the sweet tin I breathed a sigh of relief and looked around to see what had become of my dog, whose presence I had completely forgotten during the tense and exciting moments of making my first capture of an adder. There on a nearby tussock of grass he lay curled up fast asleep, completely oblivious to everything, and having apparently missed what had seemed to me at that time as a life and death struggle between his master and a dangerously venomous snake. The long, hot tramp across the heath had tired my dog out, for he was only a few months old, and like the adder he had found the springy couch of grass too comfortable to resist. It was just as well that he had fallen asleep, as I would not have liked to have had to cope with both an inquisitive young dog and a squirming, angry adder at the same time. (I feel I should add that the adder I captured that day was returned to the heath about a fortnight later and was released close to the spot where it had been caught.)

Having at last succeeded in finding and capturing my very first adder I was in an exultant and triumphant mood. Waking the dog from his nap I called him to follow. The sun had now reached the horizon, and as the dog and I made our way off the heath, curlews were lifting from the marsh and planing down again with soulful cries which seemed expressive of this wild and desolate place.

Index